UTILITY OF QUATERNIONS IN PHYSICS

UTILITY OF QUATERNIONS IN PHYSICS

Alexander McAulay M.A.,
LECTURER IN MATHEMATICS AND PHYSICS
IN THE UNIVERSITY OF TASMANIA

HAWK PRESS

Published by

Hawk Press
4836/24, Ansari Road, Daryaganj
New Delhi – 110 002
Phones : 9643330713, 011-23278618, 011-35676207
E-mail: thehawkpress@gmail.com
www.thehawkpress.com

ISBN: 978-93-93971-84-5

PREFACE.

The present publication is an essay that was sent in (December, 1887) to compete for the Smith's Prizes at Cambridge.

To the onlooker it is always a mournful thing to see what he considers splendid abilities or opportunities wasted for lack of knowledge of some paltry commonplace truth. Such is in the main my feeling when considering the neglect of the study of Quaternions by that wonderful corporation the University of Cambridge. To the alumnus she is apt to appear as the leader in all branches of Mathematics. To the outsider she appears rather as the leader in *Applied* Mathematics and as a ready welcomer of other branches.

If Quaternions were simply a branch of Pure Mathematics we could understand why the study of them was almost confined to the University which gave birth to them, but as the truth is quite otherwise it is hard to shew good reason why they have not struck root also in Cambridge. The prophet on whom Hamilton's mantle has fallen is more than a mathematician and more than a natural philosopher—he is both, and it is to be noted also that he is a Cambridge man. He has preached in season and out of season (if that were possible) that Quaternions are especially useful in Physical applications. Why then has his Alma Mater turned a deaf ear? I cannot believe that she is in her dotage and has lost her hearing. The problem is beyond me and I give it up.

But I wish to add my little efforts to Prof. Tait's powerful advocacy to bring about another state of affairs. Cambridge is the prepared ground on which if anywhere the study of the Physical applications of Quaternions ought to flourish.

When I sent in the essay I had a faint misgiving that perchance there was not a single man in Cambridge who could understand it without much labour—and yet it is a straightforward application of Hamilton's principles. I cannot say what transformation scene has taken place in the five years that have elapsed, but an encouraging fact is that one professor at any rate has been imported from Dublin.

There is no lack in Cambridge of the cultivation of Quaternions *as an algebra*, but this cultivation is not Hamiltonian, though an evidence of the great fecundity of Hamilton's work. Hamilton looked upon Quaternions as a *geometrical* method, and it is in this respect that he has as yet failed to find worthy followers resident in Cambridge. [The chapter contributed by Prof. Cayley to Prof. Tait's 3rd ed. of 'Quaternions' deals with quite a different subject from the rest of the treatise, a subject that deserves a distinctive name, say, *Cayleyan Quaternions.*]

I have delayed for a considerable time the present publication in order at the last if possible to make it more effective. I have waited till I could by a more striking example than any in the essay shew the immense utility of Quaternions in the regions in which I believe them to be especially powerful. This I believe has been done in the 'Phil. Trans.' 1892, p. 685. Certainly on two occasions copious extracts have been published, viz. in the P. R. S. E., 1890–1, p. 98, and in the 'Phil. Mag.' June 1892, p. 477, but the reasons are simple. The first was published after the subject of the 'Phil. Trans.' paper had been considered sufficiently to afford clear daylight ahead in that direction, and the second after that paper had actually been despatched for publication.

At the time of writing the essay I possessed little more than faith in the potentiality of Quaternions, and I felt that something more than faith was needed to convince scientists. It was thought that rather than publish in driblets it were better to wait for a more copious shower on the principle that a well-directed heavy blow is more effective than a long-continued series of little pushes.

Perhaps more harm has been done than by any other cause to the study of Quaternions in their Physical applications by a silly superstition with which the nurses of Cambridge are wont to frighten their too timorous charges. This is the belief that the subject of Quaternions is difficult. It is difficult in one sense and in no other, in that sense in which the subject of analytical conics is difficult to the schoolboy, in the sense in which every subject is difficult whose fundamental ideas and methods are different from any the student has hitherto been introduced to. The only way to convince the nurses that Quaternions form a healthy diet for the *young* mathematician is to prove to them that they will "pay" in the first part of the Tripos. Of course this is an impossible task while the only questions set in the Tripos on the subject are in the second part and average one in two years. [This solitary biennial question is rarely if ever anything but an exercise in *algebra.* The very form in which candidates are invited, or at any rate were in my day, to study Quaternions is an insult to the memory of Hamilton. The monstrosity "Quaternions and other non-commutative algebras" can only be paralleled by "Cartesian Geometry and other commutative algebras." When I was in Cambridge it was currently reported that if an answer to a Mathematical Tripos question were couched in Hebrew the candidate would or would not get credit for the answer according as one or more of the examiners did or did not understand Hebrew, and that in this respect Hebrew or Quaternions were strictly analogous.]

Is it hopeless to appeal to the charges? I will try. Let me suppose that some budding Cambridge Mathematician has followed me so far. I now address myself to him. Have you ever felt a joy in Mathematics? Probably you have, but it was before your schoolmasters had found you out and resolved to fashion you into an examinee. Even now you occasionally have feelings like the dimly remembered ones. Now and then you forget that you are nerving yourself for that Juggernaut the Tripos. Let me implore you as though your soul's salvation depended on it to let these trances run their utmost course in spite of solemn warnings from your nurse. You will in time be rewarded by a soul-thrilling dream whose subject is the Universe and whose organ to look upon the Universe withal is the sense called Quaternions. Steep yourself in the delirious pleasures. When you wake you will have forgotten the Tripos and in the fulness of time will develop into a financial wreck, but in possession of the memory of that heaven-sent dream you will be a far happier and richer man than the millionest millionaire.

To pass to earth—from the few papers I have published it will be evident that the subject treated of here is one I have very much at heart, and I think that the publication of the essay is likely to conduce to an acceptance of the view that *it is now the duty of mathematical physicists to study Quaternions seriously*. I have been told by more than one of the few who have read some of my papers that they prove rather stiff reading. The reasons for this are not in the papers I believe but in matters which have already been indicated. Now the present essay reproduces the order in which the subject was developed in my own mind. The less complete treatment of a subject, especially if more diffuse, is often easier to follow than the finished product. It is therefore probable that the present essay is likely to prove more easy reading than my other papers.

Moreover I wish it to be studied by a class of readers who are not in the habit of consulting the proceedings, &c., of learned societies. I want the slaves of examination to be arrested and to read, for it is apparently to the rising generation that we must look to wipe off the blot from the escutcheon of Cambridge.

And now as to the essay itself. But one real alteration has been made. A passage has been suppressed in which were made uncomplimentary remarks concerning a certain author for what the writer regards as his abuse of Quaternion methods. The author in question would no doubt have been perfectly well able to take care of himself, so that perhaps there was no very good reason for suppressing the passage as it still represents my convictions, but I did not want a side issue to be raised that would serve to distract attention from the main one. To bring the notation into harmony with my later papers dv and ∇' which occur in the manuscript have been changed throughout to $d\Sigma$ and Δ respectively. To facilitate printing the solidus has been freely introduced and the vinculum abjured. Mere slips of the pen have been corrected. A formal prefatory note required by the conditions of competition has been omitted. The Table of Contents was not prefixed to the original essay. It consists of little more than a collection of the headings scattered through the essay. Several notes have been added, all indicated by square brackets and the date (1892 or 1893). Otherwise the essay remains absolutely unaltered. The name originally given to the essay is at the head of p. 1 below. The name on the title-page is adopted to prevent confusion of the essay with the 'Phil. Mag.', paper referred to above. What in the peculiar calligraphy of the manuscript was meant for the familiar $\iiint () \, d\varsigma$ has been consistently rendered by the printer as $\iiint () \, ds$. As the mental operation of substituting the former for the latter is not laborious I have not thought it necessary to make the requisite extensive alterations in the proofs.

I wish here to express my great indebtedness to Prof. Tait, not only for having through his published works given me such knowledge of Quaternions as I possess but for giving me private encouragement at a time I sorely needed it. There was a time when I felt tempted to throw my convictions to the winds and follow the line of least resistance. To break down the solid and well-nigh universal scepticism as to the utility of Quaternions in Physics seemed too much like casting one's pearls—at least like crying in the wilderness.

But though I recognise that I am fighting under Prof. Tait's banner, yet, as every subaltern could have conducted a campaign better than his general, so in some details I feel compelled to differ from Professor Tait. Some two or three years ago he was good enough to read the present essay. He somewhat severely criticised certain points but did not convince me on all.

Among other things he pointed out that I sprung on the unsuspicious reader without due warning and explanation what may be considered as a peculiarity in symbolisation. I take this opportunity therefore of remedying the omission. In Quaternions on account of the non-commutative nature of multiplication we have not the same unlimited choice of order of the terms in a product as we have in ordinary algebra, and the same is true of certain quaternion operators. It is thus inconvenient in many cases to use the familiar method of indicating the connection between an operator and its operand by placing the former immediately before the latter. Another method is adopted. With this other method the operator may be separated from the operand, but it seems that there has been a tacit convention among users of this method that the separated operator is still to be restricted to precedence of the operand. There is of course nothing in the nature of things why this should be so, though its violation may seem a trifle strange at first, just as the tyro in Latin is puzzled by the unexpected corners of a sentence in which adjectives (operators) and their nouns (operands) turn up. Indeed a Roman may be said to have anticipated in every detail the method of indicating the connection now under discussion, for he did so by the similarity of the suffixes of his operators and operands. In this essay his example is followed and therefore no restrictions except such as result from the genius of the language (the laws of Quaternions) are placed on the relative positions in a product of operators and operands. With this warning the reader ought to find no difficulty.

One of Prof. Tait's criticisms already alluded to appears in the third edition of his 'Quaternions.' The process held up in § 500 of this edition as an example of "how not to do it" is contained in § 6 below and was first given in the 'Mess. of Math.,' 1884. He implies that the process is a "most intensely artificial application of" Quaternions. If this were true I should consider it a perfectly legitimate criticism, but I hold that it is the exact reverse of the truth. In the course of Physical investigations certain volume integrals are found to be capable of, or by general considerations are obviously capable of transformation into surface integrals. We are led to seek for the correct expression in the latter form. Starting from this we can by a long, and in my opinion, tedious process arrive at the most general type of volume integral which is capable of transformation into a surface integral. [I may remark in passing that Prof. Tait did not however arrive at quite the most general type.] Does it follow that this is the most natural course of procedure? Certainly not, as I think. It would be the most natural course for the empiricist, but not for the scientist. When he has been introduced to one or two volume integrals capable of the transformation the *natural* course of the mathematician is to ask himself what is the most general volume integral of the kind. By quite elementary considerations he sees that while only such volume integrals as satisfy certain conditions are transformable into surface integrals, yet *any* surface integral which is continuous and applies to the complete boundary of any finite volume can be expressed as a volume integral throughout that volume. He is thus led to start from the surface integral and deduces by the briefest of processes the most general volume integral of the type required. Needless to say, when giving his demonstration he does not bare his soul in this way. He thinks rightly that any mathematician can at once divine the exact road he has followed. Where is the artificiality?

Let me in conclusion say that even now I scarcely dare state what I believe to be the proper place of Quaternions in a Physical education, for fear my statements be regarded as the uninspired babblings of a misdirected enthusiast, but I cannot refrain from saying that I look forward to the time when Quaternions will appear in every Physical text-book that assumes the knowledge of (say) elementary plane trigonometry.

I am much indebted to Mr G. H. A. Wilson of Clare College, Cambridge, for helping me in the revision of the proofs, and take this opportunity of thanking him for the time and trouble he has devoted to the work.

ALEX. McAULAY.

UNIVERSITY OF TASMANIA,
HOBART.
March 26, 1893.

CONTENTS.

SECTION V HYDRODYNAMICS

SECTION VI. THE VORTEX-ATOM THEORY

QUATERNIONS AS A PRACTICAL INSTRUMENT OF PHYSICAL RESEARCH.

SECTION I.

INTRODUCTION.

It is a curious phenomenon in the History of Mathematics that the greatest work of the greatest Mathematician of the century which prides itself upon being the most enlightened the world has yet seen, has suffered the most chilling neglect.

The cause of this is not at first sight obvious. We have here little to do with the benefit provided by Quaternions to Pure Mathematics. The reason for the neglect here *may* be that Hamilton himself has developed the Science to such an extent as to make successors an impossibility. One cannot however resist a strong suspicion that were the subject even studied we should hear more from Pure Mathematicians, of Hamilton's valuable results. This reason at any rate cannot be assigned for the neglect of the Physical side of Quaternions. Hamilton has done but little in this field, and yet when we ask what Mathematical Physicists have been tempted by the bait to win easy laurels (to put the incentive on no higher grounds), the answer must be scarcely one. Prof. Tait is the grand exception to this. But well-known Physicist though he be, his fellow-workers for the most part render themselves incapable of appreciating his valuable services by studying the subject if at all only as dilettanti. The number who read a small amount in Quaternions is by no means small, but those who get further than what is recommended by Maxwell as imperatively necessary are but a small percentage of the whole.

I cannot help thinking that this state of affairs is owing chiefly to a prejudice. This prejudice is well seen in Maxwell's well-known statement—"I am convinced that the introduction of the ideas, *as distinguished from the operations and methods* of Quaternions, will be of great use to us in all parts of our subject."* Now what I hold and what the main object of this essay is to prove is that

**Elect. and Mag.* Vol. I. § 10.

the "operations and methods" of Quaternions are as much better qualified to deal with Physics than the ordinary ones as are the "ideas".

But, what has produced this notion, that the subject of Quaternions is only a pretty toy that has nothing to do with the serious work of practical Physics? It must be the fact that it has hitherto produced few results that appeal strongly to Physicists. This I acknowledge, but that the deduction is correct I strongly disbelieve. As well might an instrument of which nobody has attempted to master the principles be blamed for not being of much use. Workers naturally find themselves while still inexperienced in the use of Quaternions incapable of clearly thinking through them and of making them do the work of Cartesian Geometry, and they conclude that Quaternions do not provide suitable treatment for what they have in hand. The fact is that the subject requires a slight development in order readily to apply to the practical consideration of most physical subjects. The first steps of this, which consist chiefly in the invention of new symbols of operation and a slight examination of their chief properties, I have endeavoured to give in the following pages.

I may now state what I hold to be the mission of Quaternions to Physics. I believe that Physics would advance with both more rapid and surer strides were Quaternions introduced to serious study to the almost total exclusion of Cartesian Geometry, except in an insignificant way as a particular case of the former. All the geometrical processes occurring in Physical *theories* and *general* Physical problems are much more graceful in their Quaternion than in their Cartesian garb. To illustrate what is here meant by "theory" and "general problem" let us take the case of Elasticity treated below. That by the methods advocated not only are the already well-known results of the general *theory* of Elasticity better proved, but more general results are obtained, will I think be acknowledged after a perusal of § 12 to § 21 below. That Quaternions are superior to Cartesian Geometry in considering the *general problems* of (1) an infinite isotropic solid, (2) the torsion and bending of prisms and cylinders, (3) the general theory of wires, I have endeavoured to shew in § 22–§ 33. But for *particular* problems such as the torsion problem for a cylinder of given shape, we require of course the various theories specially constructed for the solution of particular problems such as Fourier's theories, complex variables, spherical harmonics, &c. It will thus be seen that I do not propose to banish these theories but merely Cartesian Geometry.

So mistaken are the common notions concerning the pretensions of advocates of Quaternions that I was asked by one well-known Mathematician whether Quaternions furnished methods for the solution of differential equations, as he asserted that this was all that remained for Mathematics in the domain of Physics!

Quaternions can no more solve differential equations than Cartesian Geometry, but the solution of such equations can be performed as readily, in fact generally more so, in the Quaternion shape as in the Cartesian. But that the sole work of Physical Mathematics to-day is the solution of differential equations I beg to question. There are many and important Physical questions and extensions of Physical theories that have little or nothing to do with such solutions. As witness I may call attention to the new Physical work which occurs below.

If only on account of the extreme simplicity of Quaternion notation, large advances in the parts of Physics now indicated, are to be expected. Expressions which are far too cumbrous to be of much use in the Cartesian shape become so simple when translated into Quaternions, that they admit of easy interpretation and, what is perhaps of more importance, of easy manipulation. Compare for instance the particular case of equation (15*m*) § 16 below when $\mathbf{F} = 0$ with the same thing as considered in Thomson and Tait's *Nat. Phil.*, App. C. The Quaternion equation is

$$\rho_1' S \nabla_{1} \Psi \varpi w \Delta = 0.$$

The Cartesian *exact* equivalent consists of Thomson and Tait's equations (7), viz.

$$\frac{d}{dx}\left\{2\frac{dw}{dA}\left(\frac{d\alpha}{dx}+1\right)+\frac{dw}{db}\frac{d\alpha}{dz}+\frac{dw}{dc}\frac{d\alpha}{dy}\right\}$$
$$+\frac{d}{dy}\left\{2\frac{dw}{dB}\frac{d\alpha}{dy}+\frac{dw}{da}\frac{d\alpha}{dz}+\frac{dw}{dc}\left(\frac{d\alpha}{dx}+1\right)\right\}$$
$$+\frac{d}{dz}\left\{2\frac{dw}{dC}\frac{d\alpha}{dz}+\frac{dw}{da}\frac{d\alpha}{dy}+\frac{dw}{db}\left(\frac{d\alpha}{dx}+1\right)\right\}=0,$$

and two similar equations.

Many of the equations indeed in the part of the essay where this occurs, although quite simple enough to be thoroughly useful in their present form, lead to much more complicated equations than those just given when translated into Cartesian notation.

It will thus be seen that there are two statements to make good:—(1) that Quaternions are in such a stage of development as already to justify the practically complete banishment of Cartesian Geometry from Physical questions of a general nature, and (2) that Quaternions will in Physics produce many *new* results that cannot be produced by the rival and older theory.

To establish completely the first of these propositions it would be necessary to go over *all* the ground covered by Mathematical Physical Theories, by means of our present subject, and compare the proofs with the ordinary ones. This of course is impossible in an essay. It would require a treatise of no small dimensions. But the principle can be followed to a small extent. I have therefore taken three typical theories and applied Quaternions to most of the general propositions in them. The three subjects are those of Elastic Solids, with the thermodynamic considerations necessary, Electricity and Magnetism, and Hydrodynamics. It is impossible without greatly exceeding due limits of space to consider in addition, Conduction of Heat, Acoustics, Physical Optics, and the Kinetic Theory of Gases. With the exception of the first of these subjects I do not profess even to have attempted hitherto the desired applications, but one would seem almost justified in arguing that, since Quaternions have been found so applicable to the subjects considered, they are very likely to prove useful to about the same extent in similar theories. Again, only in one of the subjects chosen, viz., Hydrodynamics, have I given the *whole* of the general theory which usually appears in text-books. For instance, in Electricity and Magnetism I have not considered Electric Conduction in three dimensions which, as Maxwell remarks, lends itself very readily to Quaternion treatment, nor Magnetic Induction, nor the Electro-Magnetic Theory of Light. Again, I have left out the consideration of Poynting's theories of Electricity which are very beautifully treated by Quaternions, and I felt much tempted to introduce some considerations in connection with the Molecular Current theory of Magnetism. With similar reluctance I have been compelled to omit many applications in the Theory of Elastic Solids, but the already too large size of the essay admitted of no other course. Notwithstanding these omissions, I think that what I have done in this part will go far to bear out the truth of the first proposition I have stated above.

But it is the second that I would especially lay stress upon. In the first it is merely stated that Cartesian Geometry is an antiquated machine that ought to be thrown aside to make room for modern improvements. But the second asserts that the improved machinery will not only do the work of the old better, but will also do much work that the old is quite incapable of doing at all. Should this be satisfactorily established and should Physicists in that case still refuse to have anything to do with Quaternions, they would place themselves in the position of the traditional workmen who so strongly objected to the introduction of machinery to supplant manual labour.

But in a few months and synchronously with the work I have already described, to arrive at a *large* number of new results is too much to expect even from such a subject as that now under discussion. There are however some few such results to shew. I have endeavoured to advance each of the theories chosen in at least one direction. In the subject of Elastic Solids I have expressed the stress in terms of the strain in the most general case, i.e. where the strain is not small, where the ordinary assumption of no stress-couple is not made and where no assumption is made as to homogeneity, isotropy, &c. I have also obtained the equations of motion when there is given an external force and couple per unit volume of the unstrained solid. These two problems, as will be seen, are by no means identical. In Electrostatics I have considered the most general mechanical results flowing from Maxwell's theory, and their explanation by stress in the dielectric. These results are not known, as might be inferred from this mode of statement, for to solve the problem we require to know forty-two independent constants to express the properties of the dielectric at a given state of strain at each point. These are the six coefficients of specific inductive capacity and their thirty-six differential coefficients with regard to the six coordinates of pure strain. But, as far as I am aware, only such particular cases of this have already been considered as make the forty-two constants reduce at most to three. In Hydrodynamics I have endeavoured to deduce certain general phenomena which would be exhibited by vortex-atoms acting upon one another. This has been done by examination of an equation which has not, I believe, been hitherto given. The result of this part of the essay is to lead to a presumption against Sir William Thomson's Vortex-Atom Theory and in favour of Hicks's.

As one of the objects of this introduction is to give a bird's-eye view of the merits of Quaternions as opposed to Cartesian Geometry, it will not be out of place to give side by side the Quaternion and the Cartesian forms of most of the new results I have been speaking about. It must be premised, as already hinted, that the usefulness of these results must be judged not by the Cartesian but by the Quaternion form.

Elasticity.

Let the point (x, y, z) of an elastic solid be displaced to (x', y', z'). The strain at any point that is caused may be supposed due to a pure strain followed by a rotation. In Section III. below, this pure strain is called ψ. Let its coordinates be e, f, g, $a/2$, $b/2$, $c/2$; i.e. if the vector (ξ, η, ζ) becomes (ξ', η', ζ') by means of the pure strain, then

$$\xi' = e\xi + \tfrac{1}{2}c\eta + \tfrac{1}{2}b\zeta,$$
$$\text{\&c., \&c.}$$

Thus when the strain is small e, f, g reduce to Thomson and Tait's $1 + e$, $1 + f$, $1 + g$ and a, b, c are the same both in their case and the present one. Now let the coordinates of Ψ, § 16 below, be E, F, G, $A/2$, $B/2$, $C/2$. Equation (15), § 16 below, viz.

$$^*\Psi\omega = \psi^2\omega = \chi'\chi\omega = \nabla_1 S\rho_1'\rho_2' S\omega\nabla_2,$$

gives in our present notation

$$E = e^2 + c^2/4 + b^2/4 = (dx'/dx)^2 + (dy'/dx)^2 + (dz'/dx)^2,$$
$$\&c., \&c.$$
$$A = a(f+g) + bc/2$$
$$= 2\{(dx'/dy)(dx'/dz) + (dy'/dy)(dy'/dz) + (dz'/dy)(dz'/dz)\},$$
$$\&c., \&c.$$

which shew that the present E, F, G, $A/2$, $B/2$, $C/2$ are the A, B, C, a, b, c of Thomson and Tait's *Nat. Phil.*, App. C.

Let us put

$$J\begin{pmatrix} x'y'z' \\ xyz \end{pmatrix} = J$$
$$J\begin{pmatrix} y'z' \\ yz \end{pmatrix} = J_{11}, \quad \&c., \&c.,$$
$$J\begin{pmatrix} z'x' \\ yz \end{pmatrix} = J_{12}, J\begin{pmatrix} y'z' \\ zx \end{pmatrix} = J_{21}, \quad \&c., \&c., \&c., \&c.$$

I have shewn in § 14 below that the *stress-couple is quite independent of the strain.* Thus we may consider the stress to consist of two parts—an ordinary stress $PQRSTU$ as in Thomson and Tait's *Nat. Phil.* and a stress which causes a couple per unit volume $L'M'N'$. The former only of these will depend on strain. The result of the two will be to cause a force (as indeed can be seen from the expressions in § 13 below) per unit area on the x-interface P, $U + N'/2$, $T - M'/2$, and so for the other interfaces. If L, M, N be the external couple per unit volume of the *unstrained* solid we shall have

$$L' = -L/J, \quad M' = -M/J, \quad N' = -N/J,$$

*This result is one of Tait's (*Quaternions* § 365 where he has $\phi'\phi = \varpi^2$). It is given here for completeness.

for the external couple and the stress-couple are *always* equal and opposite. Thus the force on the x-interface becomes

$$P, \quad U - N/2J, \quad T + M/2J$$

and similarly for the other interfaces.

To express the part of the stress (P &c.) which depends on the strain in terms of that strain, consider w the potential energy per unit volume of the unstrained solid as a function of E &c. In the general thermodynamic case w may be defined by saying that

$w \times$ (the element of volume)
= (the intrinsic energy of the element)
− (the entropy of the element × its absolute temperature × Joule's coefficient).

Of course w may be, and indeed is in § 14, § 15 below, regarded as a function of e &c.

The equation for stress is (15b) § 16 below, viz.,

$$J\bar{\phi}\omega = 2\chi\Psi\sqsupset w\chi'\omega = 2\rho_1' S\rho_2'\omega S\nabla_1\Psi\sqsupset w\nabla_2.$$

The second of the expressions is in terms of the strain and the third in terms of the displacement and its derivatives. In our present notation this last is

$$\frac{JP}{2} = \left(\frac{dx'}{dx}\right)^2 \frac{dw}{dE} + \left(\frac{dx'}{dy}\right)^2 \frac{dw}{dF} + \left(\frac{dx'}{dz}\right)^2 \frac{dw}{dG}$$
$$+ 2\frac{dx'}{dy}\frac{dx'}{dz}\frac{dw}{dA} + 2\frac{dx'}{dz}\frac{dx'}{dx}\frac{dw}{dB} + 2\frac{dx'}{dx}\frac{dx'}{dy}\frac{dw}{dC},$$

&c., &c.

$$\frac{JS}{2} = \frac{dy'}{dx}\frac{dz'}{dx}\frac{dw}{dE} + \frac{dy'}{dy}\frac{dz'}{dy}\frac{dw}{dF} + \frac{dy'}{dz}\frac{dz'}{dz}\frac{dw}{dG}$$
$$+ \left(\frac{dy'}{dy}\frac{dz'}{dz} + \frac{dy'}{dz}\frac{dz'}{dy}\right)\frac{dw}{dA} + \left(\frac{dy'}{dz}\frac{dz'}{dx} + \frac{dy'}{dx}\frac{dz'}{dz}\right)\frac{dw}{dB}$$
$$+ \left(\frac{dy'}{dx}\frac{dz'}{dy} + \frac{dy'}{dy}\frac{dz'}{dx}\right)\frac{dw}{dC},$$

&c., &c.

In § 14 I also obtain this part of the stress explicitly in terms of e, f, g, a, b, c, of w as a function of these quantities and of the axis and amount of rotation. But these results are so very complicated in their Cartesian shape that it is quite useless to give them.

To put down the equations of motion let X_x, Y_x, Z_x be the force due to stress on what before strain was unit area perpendicular to the axis of x. Similarly for X_y, &c. Next suppose that X, Y, Z is the external force per unit volume of the unstrained solid and let D be the original density of the solid. Then the equation of motion (15n) § 16a below, viz.

$$D\ddot{\rho}' = \mathbf{F} + \tau\Delta,$$

gives in our present notation

$$X + dX_x/dx + dX_y/dy + dX_z/dz = \ddot{x}'D, \quad \text{\&c., \&c.}$$

It remains to express X_x &c. in terms of the displacement and LMN. This is done in equation (15l) § 16 below, viz.

$$\tau\omega = -2\rho_1' S\nabla_1 \psi\text{ɑ} w\omega + 3V\mathbf{M}V\rho_1'\rho_2' S\,\omega\nabla_1\nabla_2/2S\nabla_1\nabla_2\nabla_3 S\rho_1'\rho_2'\rho_3'.^*$$

In our present notation this consists of the following nine equations:

$$X_x = 2\left(\frac{dw}{dE}\frac{dx'}{dx} + \frac{dw}{dC}\frac{dx'}{dy} + \frac{dw}{dB}\frac{dx'}{dz}\right) + \frac{J_{12}N - J_{13}M}{2J},$$

$$Y_x = 2\left(\frac{dw}{dE}\frac{dy'}{dx} + \frac{dw}{dC}\frac{dy'}{dy} + \frac{dw}{dB}\frac{dy'}{dz}\right) + \frac{J_{13}L - J_{11}N}{2J},$$

$$Z_x = 2\left(\frac{dw}{dE}\frac{dz'}{dx} + \frac{dw}{dC}\frac{dz'}{dy} + \frac{dw}{dB}\frac{dz'}{dz}\right) + \frac{J_{11}M - J_{12}L}{2J},$$

and six similar equations.

We thus see that in the case where LMN are zero, our present X_x, X_y, X_z are the PQR of Thomson and Tait's *Nat. Phil.* App. C (d), and therefore equations (7) of that article agree with our equations of motion when we put both the external force and the acceleration zero.

---

*The second term on the right contains *in full* the nine terms corresponding to $(J_{12}N - J_{13}M)/2J$. Quaternion notation is therefore here, as in nearly all cases which occur in Physics, considerably more compact even than the notations of determinants or Jacobians.

These are some of the new results in Elasticity, but, as I have hinted, there are others in § 14, § 15 which it would be waste of time to give in their Cartesian form.

Electricity.

In Section IV. below I have considered, as already stated, the most general mechanical results flowing from Maxwell's theory of Electrostatics. I have shewn that here, as in the particular cases considered by others, the forces, whether per unit volume or per unit surface, can be explained by a stress in the dielectric. It is easiest to describe these forces by means of the stress.

Let the coordinates of the stress be $PQRSTU$. Then $F_1F_2F_3$ the mechanical force, due to the field per unit volume, exerted upon the dielectric where there is no discontinuity in the stress, is given by

$$F_1 = dP/dx + dU/dy + dT/dz, \quad \&c., \&c.$$

and (l, m, n) being the direction cosines of the normal to any surface, pointing away from the region considered

$$F_1' = -[lP + mU + nT]_a - [\ ]_b, \quad \&c., \&c.,$$

where a, b indicate the two sides of the surface and F_1', F_2', F_3' is the force due to the field per unit surface.

It remains to find P &c. Let X, Y, Z be the electro-motive force, α, β, γ the displacement, w the potential energy per unit volume and K_{xx}, K_{yy}, K_{zz}, K_{yz}, K_{zx}, K_{xy} the coefficients of specific inductive capacity. Let $1+e$, $1+f$, $1+g$, $a/2$, $b/2$, $c/2$ denote the pure part of the strain of the medium. The K's will then be functions of e &c. and we must suppose these functions known, or at any rate we must assume the knowledge of both the values of the K's and their differential coefficients at the particular state of strain in which the medium is when under consideration. The relations between the above quantities are

$$4\pi\alpha = K_{xx}X + K_{xy}Y + K_{zx}Z, \quad \&c., \&c.$$

$$\begin{aligned} w &= (X\alpha + Y\beta + Z\gamma)/2 \\ &= (K_{xx}X^2 + K_{yy}Y^2 + K_{zz}Z^2 + 2K_{yz}YZ + 2K_{zx}ZX + 2K_{xy}XY)/8\pi. \end{aligned}$$

It is the second of these expressions for w which is assumed below, and the differentiations of course refer only to the K's. The equation expressing P &c. in terms of the field is (21) § 40 below, viz.

$$\phi\omega = -\tfrac{1}{2}V\mathbf{D}\omega\mathbf{E} - {}_{\Psi}\daleth w\,\omega,$$

which in our present notation gives the following six equations

$$P = -\tfrac{1}{2}(-\alpha X + \beta Y + \gamma Z) - dw/de, \quad \&c., \&c.,$$
$$S = \tfrac{1}{2}(\beta Z + \gamma Y) - dw/da, \quad \&c., \&c.$$

I have shewn in § 41–§ 45 below that these results agree with particular results obtained by others.

Hydrodynamics.

The new work in this subject is given in Section VI.—"The Vortex-Atom Theory." It is quite unnecessary to translate the various expressions there used into the Cartesian form. I give here only the principal equation in its two chief forms, equation (9) § 89 and equation (11) § 90, viz.

$$P + v - \sigma^2/2 + (4\pi)^{-1}\iiint(S\sigma\tau\nabla u + u\partial m/\partial t)\,d\mathfrak{s} = H,$$
$$P + v - \sigma^2/2 + (4\pi)^{-1}\iiint\{d\mathfrak{s}S\,\nabla u(V\sigma\tau - m\sigma) + ud(md\mathfrak{s})/dt\} = H.$$

In Cartesian notation these are

$$\int dp/\rho + V + q^2/2$$
$$- (4\pi)^{-1}\iiint\{2[(x' - x)(w\eta - v\zeta) + \cdots + \cdots]/r^3$$
$$+ (\partial c/\partial t)/r\}\,dx'\,dy'\,dz' = H.$$

$$\int dp/\rho + V + q^2/2$$
$$- (4\pi)^{-1}\iiint\{(x' - x)[2(w\eta - v\zeta) - cu] + \cdots + \cdots\}/r^3.\,dx'\,dy'\,dz'$$
$$- (4\pi)^{-1}\iiint\{d(c\,dx'\,dy'\,dz')/dt\}/r = H.$$

The fluid here considered is one whose motion is continuous from point to point and which extends to infinity. The volume integral extends throughout space. The notation is as usual. It is only necessary to say that H is a function of the time only, r is the distance between the points x', y', z' and x, y, z;

$$c = du/dx + dv/dy + dw/dz;$$

d/dt is put for differentiation which follows a particle of the fluid, and $\partial/\partial t$ for that which refers to a fixed point.

The explanation of the unusual length of this essay, which I feel is called for, is contained in the foregoing description of its objects. If the objects be justifiable, so must also be the length which is a necessary outcome of those objects.

SECTION II.

QUATERNION THEOREMS.

Definitions.

1. As there are two or three symbols and terms which will be in constant use in the following pages that are new or more general in their signification than is usual, it is necessary to be perhaps somewhat tediously minute in a few preliminary definitions and explanations.

A *function* of a variable in the following essay is to be understood to mean anything which depends on the variable and which can be subjected to mathematical operations; the variable itself being anything capable of being represented by a mathematical symbol. In Cartesian Geometry the variable is generally a single scalar. In Quaternions on the other hand a general quaternion variable is not infrequent, a variable which requires 4 scalars for its specification, and similarly for the function. In both, however, either the variable or the function may be a mere symbol of operation. In the following essay we shall frequently have to speak of variables and functions which are neither quaternions nor mere symbols of operation. For instance K in § 40 below requires 6 scalars to specify it, and it is a function of ψ which requires 6 scalars and ρ which requires 3 scalars. When in future the expression "any function" is used it is always to be understood in the general sense just explained.

We shall frequently have to deal with functions of many independent vectors, and especially with functions which are linear in each of the constituent vectors. These functions merely require to be noticed but not defined.

Hamilton has defined the meaning of the symbolic vector ∇ thus:—

$$\nabla = i\frac{d}{dx} + j\frac{d}{dy} + k\frac{d}{dz},$$

where i, j, k are unit vectors in the directions of the mutually perpendicular axes x, y, z. I have found it necessary somewhat to expand the meaning of this symbol. When a numerical suffix 1, 2, ... is attached to a ∇ in any expression it is to indicate that the differentiations implied in the ∇ are to refer to and only to other symbols in the same expression which have the same suffix. After the implied differentiations have been performed the suffixes are of course removed. Thus $Q(\alpha, \beta, \gamma, \delta)$ being a quaternion function of any four vectors α, β, γ, δ, linear in each

$$\begin{aligned} Q(\lambda_1\mu_2\nabla_1\nabla_2) \equiv\ & Q\left(\frac{d\lambda}{dx}\frac{d\mu}{dx}ii\right) + Q\left(\frac{d\lambda}{dy}\frac{d\mu}{dx}ji\right) + Q\left(\frac{d\lambda}{dz}\frac{d\mu}{dx}ki\right) \\ &+ Q\left(\frac{d\lambda}{dx}\frac{d\mu}{dy}ij\right) + Q\left(\frac{d\lambda}{dy}\frac{d\mu}{dy}jj\right) + Q\left(\frac{d\lambda}{dz}\frac{d\mu}{dy}kj\right) \\ &+ Q\left(\frac{d\lambda}{dx}\frac{d\mu}{dz}ik\right) + Q\left(\frac{d\lambda}{dy}\frac{d\mu}{dz}jk\right) + Q\left(\frac{d\lambda}{dz}\frac{d\mu}{dz}kk\right) \end{aligned}$$

and again

$$Q_1(\lambda_1, \mu_2, \nabla_1, \nabla_2) \equiv Q_3(\lambda_1, \mu_2, \nabla_1 + \nabla_3, \nabla_2).$$

It is convenient to reserve the symbol Δ for a special meaning. It is to be regarded as a particular form of ∇, but its differentiations are to refer to *all* the variables in the term in which it appears. Thus Q being as before

$$Q(\lambda_1, \mu, \Delta, \nabla_1) = Q_2(\lambda_1, \mu_2, \nabla_1 + \nabla_2, \nabla_1)^*.$$

If in a linear expression or function ∇_1 and ρ_1 (ρ being as usual $\equiv ix + jy + kz$) occur once each they can be interchanged. Similarly for ∇_2 and ρ_2. So often does this occur that I have thought it advisable to use a separate symbol ζ_1 for each of the two ∇_1 and ρ_1, ζ_2 for each of the two ∇_2 and ρ_2 and so for ζ_3, &c. If only one such pair occur there is of course no need for the suffix attached to ζ. Thus ζ may be looked upon as a symbolic vector or as a single term put down instead of three. For $Q(\alpha, \beta)$ being linear in each of the vectors α, β

$$Q(\zeta, \zeta) = Q(\nabla_1, \rho_1) = Q(i, i) + Q(j, j) + Q(k, k). \tag{1}$$

There is one more extension of the meaning of ∇ to be given. u, v, w being the rectangular coordinates of any vector σ, ${}_\sigma\nabla$ is defined by the equation

$${}_\sigma\nabla = i\frac{d}{du} + j\frac{d}{dv} + k\frac{d}{dw}.$$

*These meanings for $\nabla_1, \nabla_2 \ldots \Delta$ I used in a paper on "Some General Theorems in Quaternion Integration," in the *Mess. of Math.* Vol. XIV. (1884), p. 26. The investigations there given are for the most part incorporated below. [Note added, 1892, see preface as to the alteration of ∇' into Δ.]

To $_{\sigma}\nabla$ of course are to be attached, when necessary, the suffixes above explained in connection with ∇. Moreover just as for ∇_1, ρ_1 we may put ζ, ζ so also for $_{\sigma}\nabla_1$, σ_1 may we put the same.

With these meanings one important result follows at once. *The ∇_1's, ∇_2's, &c., obey all the laws of ordinary vectors whether with regard to multiplication or addition*, for the coordinates d/dx, d/dy, d/dz of any ∇ obey with the coordinates of any vector or any other ∇ all the laws of common algebra.

Just as $_{\sigma}\nabla$ may be defined as a symbolic vector whose coordinates are d/du, d/dv, d/dw so ϕ being a linear vector function of any vector whose coordinates are

$$(a_1b_1c_1\ a_2b_2c_2\ a_3b_3c_3) \quad (\text{i.e. } \phi i = a_1 i + b_1 j + c_1 k, \quad \text{\&c.}).$$

$_{\phi}\text{ᗡ}$[*] is defined as a symbolic linear vector function whose coordinates are

$$(d/da_1,\ d/db_1,\ d/dc_1,\ d/da_2,\ d/db_2,\ d/dc_2,\ d/da_3,\ d/db_3,\ d/dc_3),$$

and to $_{\phi}\text{ᗡ}$ is to be applied exactly the same system of suffixes as in the case of ∇. Thus q being any quaternion function of ϕ, and ω any vector

$$\begin{aligned} {}_{\phi}\text{ᗡ}_1\omega \,.\, q_1 = &-(idq/da_1 + jdq/db_1 + kdq/dc_1)S\,i\omega \\ &-(idq/da_2 + jdq/db_2 + kdq/dc_2)S\,j\omega \\ &-(idq/da_3 + jdq/db_3 + kdq/dc_3)S\,k\omega. \end{aligned}$$

The same symbol $_{\phi}\text{ᗡ}$ is used without any inconvenience with a slightly different meaning. If the independent variable ϕ be a *self-conjugate* linear vector function it has only six coordinates. If these are $PQRSTU$ (i.e. $\phi i = Pi + Uj + Tk$, &c.) $_{\phi}\text{ᗡ}$ is defined as a self-conjugate linear vector function whose coordinates are

$$(d/dP,\ d/dQ,\ d/dR,\ \tfrac{1}{2}d/dS,\ \tfrac{1}{2}d/dT,\ \tfrac{1}{2}d/dU).$$

We shall frequently have to compare volume integrals with integrals taken over the bounding surface of the volume, and again surface integrals with integrals taken round the boundary of the surface. For this purpose we shall use the following notations for linear, surface and volume integrals respectively $\int Q\,d\rho$, $\iint Q\,d\Sigma$, $\iiint Q\,d\mathfrak{s}$ where Q is any function of the position of a point. Here $d\rho$ is a vector element of the curve, $d\Sigma$ a vector element of the surface, and $d\mathfrak{s}$ an element of volume. When comparisons between line and surface integrals are made we take $d\Sigma$ in such a direction that $d\rho$ is in the direction of positive rotation round the element $d\Sigma$ close to it. When comparisons between surface and volume integrals are made $d\Sigma$ is always taken in the direction *away from* the volume which it bounds.

[*] I have used an inverted D to indicate the analogy to Hamilton's inverted Δ.

Properties of ζ.

2. The property of ζ on which nearly all its usefulness depends is that if σ be any vector

$$\sigma = -\zeta S \zeta\sigma,$$

which is given at once by equation (1) of last section.

This gives a useful expression for the conjugate of a linear vector function of a vector. Let ϕ be the function and ω, τ any two vectors. Then ϕ' denoting as usual the conjugate of ϕ we have

$$S\omega\phi\tau = S\tau\phi'\omega,$$

whence putting on the left $\tau = -\zeta S\zeta\tau$ we have

$$S\tau(-\zeta S\omega\phi\zeta) = S\tau\phi'\omega,$$

or since τ is quite arbitrary

$$\phi'\omega = -\zeta S\omega\phi\zeta. \tag{2}$$

From this we at once deduce expressions for the pure part $\overline{\phi}\omega$ and the rotational part $V\epsilon\omega$ of $\phi\omega$ by putting

$$\left.\begin{aligned}(\phi + \phi')\omega &= -\phi\zeta S\omega\zeta - \zeta S\omega\phi\zeta = 2\overline{\phi}\omega,\\ (\phi - \phi')\omega &= -\phi\zeta S\omega\zeta - \zeta S\omega\phi\zeta = VV\zeta\phi\zeta.\omega = 2V\epsilon\omega.\end{aligned}\right\} \tag{3}$$

And all the other well-known relations between ϕ and ϕ' are at once given e.g. $S\zeta\phi\zeta = S\zeta\phi'\zeta$, i.e. the "convergence" of ϕ = the "convergence" of ϕ'.

3. Let $Q(\lambda, \mu)$ be any function of two vectors which is linear in each. Then if $\phi\omega$ be any linear vector function of a vector ω given by

$$\left.\begin{aligned}&\phi\omega = -\Sigma\beta S\omega\alpha\\ \text{we have}\quad &Q(\zeta, \phi\zeta) = \Sigma Q(-\zeta S\zeta\alpha, \beta) = \Sigma Q(\alpha, \beta),\end{aligned}\right\} \tag{4}$$

or more generally

$$Q(\zeta, \phi\chi\zeta) = \Sigma Q(\chi'\alpha, \beta). \tag{4a}$$

To prove, it is only necessary to observe that

$$\phi\chi\zeta = -\Sigma\beta S\alpha\chi\zeta = -\Sigma\beta S\zeta\chi'\alpha,$$

and that

$$-\zeta S\zeta\chi'\alpha = \chi'\alpha.$$

As a particular case of eq. (4) let ϕ have the self-conjugate value

$$\left.\begin{aligned} \phi\omega &= -\tfrac{1}{2}\Sigma(\beta S\,\omega\alpha + \alpha S\,\omega\beta). \\ \text{Then}\qquad Q(\zeta,\phi\zeta) &= \tfrac{1}{2}\Sigma\{Q(\alpha,\beta) + Q(\beta,\alpha)\}, \end{aligned}\right\} \qquad (5)$$

or if $Q(\lambda,\mu)$ is symmetrical in λ and μ

$$Q(\zeta,\phi\zeta) = \Sigma Q(\alpha,\beta). \qquad (6)$$

The application we shall frequently make of this is to the case when for α we put ∇_1 and for β, σ_1, where σ is any vector function of the position of a point. In this case the first expression for ϕ is the strain function and the second expression the pure strain function resulting from a small displacement σ at every point. As a simple particular case put $Q(\lambda,\mu) = S\,\lambda\mu$ so that Q is symmetrical in λ and μ. Thus ϕ being either of these functions

$$S\,\zeta\phi\zeta = S\,\nabla\sigma.$$

Another important equation is

$$^{*}Q(\zeta,\phi\zeta) = Q(\phi'\zeta,\zeta). \qquad (6a)$$

This is quite independent of the form of ϕ. To prove, observe that by equation (2)

$$\phi'\zeta = -\zeta_1 S\,\zeta\phi\zeta_1,$$

and that $-\zeta S\,\zeta\phi\zeta_1 = \phi\zeta_1$. Thus we get rid of ζ and may now drop the suffix of ζ_1 and so get eq. (6*a*). [Notice that by means of (6*a*), (4*a*) may be deduced from (4); for by (6*a*)

$$Q(\zeta,\phi\chi\zeta) = Q(\chi'\zeta,\phi\zeta) = \Sigma Q(\chi'\alpha,\beta) \quad \text{by (4).]}$$

3a. A more important result is the expression for $\phi^{-1}\omega$ in terms of ϕ. We assume that

$$S\,\phi\lambda\,\phi\mu\,\phi\nu = mS\,\lambda\mu\nu,$$

where λ, μ, ν are any three vectors and m is a scalar independent of these vectors. Substituting ζ_1, ζ_2, ζ_3 for λ, μ, ν and multiplying by $S\,\zeta_1\zeta_2\zeta_3$

$$S\,\zeta_1\,\zeta_2\,\zeta_3\,S\,\phi\zeta_1\,\phi\zeta_2\,\phi\zeta_3 = 6m, \qquad (6b)$$

*[Note added, 1892. For practice it is convenient to remember this in words:—*A term in which ζ and $\phi\zeta$ occur is unaltered in value by changing them into $\phi'\zeta$ and ζ respectively.*]

which gives m in terms of ϕ. That $S\zeta_1\zeta_2\zeta_3 S\zeta_1\zeta_2\zeta_3 = 6$ is seen by getting rid of each pair of ζ's in succession thus:—

$$S\zeta_1(S\zeta_1 V\zeta_2\zeta_3)\zeta_2\zeta_3 = -SV\zeta_2\zeta_3 \,.\, \zeta_2\zeta_3 = S(\zeta_2^2\zeta_3 - \zeta_2 S\zeta_2\zeta_3)\zeta_3 = -2\zeta_3\zeta_3 = 6.$$

Next observe that

$$S\phi\omega\phi\zeta_1\phi\zeta_2 = mS\omega\zeta_1\zeta_2.$$

Multiplying by $V\zeta_1\zeta_2$ and again on the right getting rid of the ζs we have

$$V\zeta_1\zeta_2 S\phi\omega\phi\zeta_1\phi\zeta_2 = -2m\omega, \tag{6c}$$

whence from equation (6*b*)

$$\omega S\zeta_1\zeta_2\zeta_3 S\phi\zeta_1\phi\zeta_2\phi\zeta_3 = -3V\zeta_1\zeta_2 S\phi\omega\phi\zeta_1\phi\zeta_2,$$

or changing ω into $\phi^{-1}\omega$

$$\phi^{-1}\omega = -\frac{3V\zeta_1\zeta_2 S\omega\phi\zeta_1\phi\zeta_2}{S\zeta_1\zeta_2\zeta_3 S\phi\zeta_1\phi\zeta_2\phi\zeta_3}. \tag{6d}$$

By equation (6*a*) of last section we can also put this in the form

$$\phi^{-1}\omega = -\frac{3V\phi'\zeta_1\phi'\zeta_2 S\omega\zeta_1\zeta_2}{S\zeta_1\zeta_2\zeta_3 S\phi'\zeta_1\phi'\zeta_2\phi'\zeta_3}, \tag{6e}$$

so that $\phi^{-1}\omega$ is obtained explicitly in terms of ϕ or ϕ'.

Equation (6*c*) or (6*d*) can be put in another useful form which is more analogous to the ordinary cubic and can be easily deduced therefrom, or *less easily from (6*d*), viz.

$$\begin{aligned} S\zeta_1\zeta_2\zeta_3(\phi^3\omega S\zeta_1\zeta_2\zeta_3 &- 3\phi^2\omega S\zeta_1\zeta_2\phi\zeta_3 \\ &+ 3\phi\omega S\zeta_1\phi\zeta_2\phi\zeta_3 - \omega S\phi\zeta_1\phi\zeta_2\phi\zeta_3) = 0. \end{aligned} \tag{6f}$$

As a useful particular case of equation (6*d*) we may notice that by equation (4) § 3 if

$$\phi\omega = -\sigma_1 S\omega\nabla_1,$$

$$\phi^{-1}\omega = -3V\nabla_1\nabla_2 S\omega\sigma_1\sigma_2 / S\nabla_1\nabla_2\nabla_3 S\sigma_1\sigma_2\sigma_3, \tag{6g}$$

and

$$\phi'^{-1}\omega = -3V\sigma_1\sigma_2 S\omega\nabla_1\nabla_2 / S\nabla_1\nabla_2\nabla_3 S\sigma_1\sigma_2\sigma_3. \tag{6h}$$

*[Note added, 1892. The cubic may be obtained in a more useful form from the equation $\omega S\zeta_1\zeta_2\zeta_3 S\phi\zeta_1\phi\zeta_2\phi\zeta_3 = -3V\zeta_1\zeta_2 S\phi\omega\phi\zeta_1\phi\zeta_2$ thus

$$\begin{aligned} V\zeta_1\zeta_2 S\phi\omega\phi\zeta_1\phi\zeta_2 &= \phi\omega S\,.\,\phi\zeta_1\phi\zeta_2 V\zeta_1\zeta_2 - \phi\zeta_1 S\,.\,\phi\omega\phi\zeta_2 V\zeta_1\zeta_2 + \phi\zeta_2 S\,.\,\phi\omega\phi\zeta_1 V\zeta_1\zeta_2 \\ &= \phi\omega S\,.\,\phi\zeta_1\phi\zeta_2 V\zeta_1\zeta_2 - 2\phi\zeta_1 S\,.\,\phi\omega\phi\zeta_2 V\zeta_1\zeta_2. \end{aligned}$$

4. Let ϕ, ψ be two linear vector functions of a vector. Then if

$$S\chi\zeta\,\phi\zeta = S\chi\zeta\,\psi\zeta,$$

where χ is a *quite arbitrary* linear vector function

$$\phi \equiv \psi,$$

for we may put $\chi\zeta = \tau S\zeta\omega$ where τ and ω are arbitrary vectors, so that

$$S\tau\,\phi\omega = S\tau\,\psi\omega,$$

or

$$\phi\omega = \psi\omega.$$

Similarly* if ϕ and ψ are both self-conjugate and χ is a quite arbitrary self-conjugate linear vector function the same relation holds as can be seen by putting

$$\chi\zeta = \tau S\zeta\omega + \omega S\zeta\tau.$$

Fundamental Property of $\text{\sf C}$.

5. Just as the fundamental property of ${}_\sigma\nabla$ is that, Q being any function of σ

$$\delta Q = -Q_1 S\,\delta\sigma\,{}_\sigma\nabla_1,$$

so we have a similar property of $\text{\sf C}$. Q being any function of ϕ a linear vector function

$$\delta Q = -Q_1 S\,\delta\phi\zeta\,{}_\phi\text{\sf C}_1\zeta. \tag{7}$$

Again $\phi\zeta_1 S\,.\,\phi\omega\,\phi\zeta_2\,V\zeta_1\zeta_2 = \phi\zeta_1 S\,.\,\phi\omega V\,.\,\phi\zeta_2\,V\zeta_1\zeta_2$

$$= \phi\zeta_1\,S\phi\omega(-\zeta_1 S\zeta_2\,\phi\zeta_2 + \zeta_2\,S\zeta_1\,\phi\zeta_2)$$

$$= -\phi(\zeta_1\,S\zeta_1\,\phi\omega)S\zeta_2\,\phi\zeta_2 + \phi(\zeta_1\,S\zeta_1\,\phi\zeta_2)\,S\zeta_2\,\phi\omega$$

$$= \phi^2\omega\,S\zeta\,\phi\zeta - \phi^2\zeta\,S\zeta\,\phi\omega = \phi^2\omega\,S\zeta\,\phi\zeta + \phi^3\omega.$$

Hence

$$\phi^3\omega - m''\phi^2\omega + m'\phi\omega - m\omega = 0,$$

where

$$6m = S\zeta_1\,\zeta_2\,\zeta_3\,S\phi\zeta_1\,\phi\zeta_2\,\phi\zeta_3$$

$$2m' = -S\,V\zeta_1\,\zeta_2\,V\phi\zeta_1\,\phi\zeta_2$$

$$m'' = -S\zeta\,\phi\zeta.]$$

*[Note added, 1892. The following slightly more general statement is a practically much more convenient form of enunciation: if $S\chi\zeta\,\phi\zeta = S\chi\zeta\,\psi\zeta$, where χ is a perfectly arbitrary self-conjugate and ϕ and ψ are not necessarily self-conjugate then $\overline{\phi} = \overline{\psi}$].

The property is proved in the same way as for ∇, viz. by expanding $S\,\delta\phi\,\zeta_\phi\daleth_1\zeta$ in terms of the coordinates of ${}_\phi\daleth$. First let ϕ be not self-conjugate, and let its nine coordinates be

$$(a_1 b_1 c_1\; a_2 b_2 c_2\; a_3 b_3 c_3).$$

Thus

$$\begin{aligned} -Q_1\,S\,\delta\phi\,\zeta\,{}_\phi\daleth_1\zeta &= -Q_1\,S\,\delta\phi i\,{}_\phi\daleth_1 i - Q_1\,S\,\delta\phi j\,{}_\phi\daleth_1 j - Q_1\,S\,\delta\phi k\,{}_\phi\daleth_1 k,\\ &= \delta a_1\, dQ/da_1 + \delta b_1\, dQ/db_1 + \delta c_1\, dQ/dc_1,\\ &+ \delta a_2\, dQ/da_2 + \delta b_2\, dQ/db_2 + \delta c_2\, dQ/dc_2,\\ &+ \delta a_3\, dQ/da_3 + \delta b_3\, dQ/db_3 + \delta c_3\, dQ/dc_3 = \delta Q. \end{aligned}$$

The proposition is exactly similarly proved when ϕ is self-conjugate.

Theorems in Integration.

6. Referring back to § 1 above for our notation for linear surface and volume integrals we will now prove that if Q be *any* linear function of a vector*

$$\int Q\,d\rho = \iint Q\,(V\,d\Sigma\,\Delta), \tag{8}$$

$$\iint Q\,d\Sigma = \iiint Q\Delta\,ds. \tag{9}$$

To prove the first divide the surface up into a series of elementary parallelograms by two families of lines—one or more members of one family coinciding with the given boundary,—apply the line integral to the boundary of each parallelogram and sum for the whole. The result will be the linear integral given in equation (8). Let the sides of one such parallelogram taken in order in the positive direction be $\alpha, \beta+\beta', -\alpha-\alpha', -\beta$; so that α' and β' are infinitely small compared with α and β, and we have the identical relation

$$0 = \alpha + \beta + \beta' - \alpha - \alpha' - \beta = \beta' - \alpha'.$$

The terms contributed to $\int Q\,d\rho$ by the sides α and $-\alpha-\alpha'$ will be (neglecting terms of the third and higher orders of small quantities)

$$Q\alpha - Q\alpha - Q\alpha' + Q_1\alpha\,S\beta\nabla_1 = -Q\alpha' + Q_1\alpha\,S\beta\nabla_1.$$

*These two propositions are generalisations of what Tait and Hicks have from time to time proved. They were first given in the present form by me in the article already referred to in § 1 above. In that paper the necessary references are given.

Similarly the terms given by the other two sides will be

$$Q\beta' - Q_1\beta\, S\,\alpha\nabla_1$$

so that remembering that $\beta' - \alpha' = 0$ and therefore $Q\beta' - Q\alpha' = 0$ we have for the whole boundary of the parallelogram

$$Q_1\alpha\, S\beta\nabla_1 - Q_1\beta\, S\,\alpha\nabla_1 = Q_1(VV\alpha\beta\,.\,\nabla_1) = QV\,d\Sigma\Delta,$$

where $d\Sigma$ is put for $V\alpha\beta$. Adding for the whole surface we get equation (8).

Equation (9) is proved in an exactly similar way by splitting the volume up into elementary parallelepipeda by three families of surfaces one or more members of one of the families coinciding with the given boundary. If α, β, γ be the vector edges of one such parallelepiped we get a term corresponding to $Q\beta' - Q\alpha'$ viz.

$$Q(\text{vector sum of surface of parallelepiped}) \equiv 0,$$

and we get the sum of three terms corresponding to

$$Q_1\alpha\, S\beta\nabla_1 - Q_1\beta\, S\,\alpha\nabla_1$$

above, viz.

$$-Q_1(V\beta\gamma)\,S\,\alpha\nabla_1 - Q_1(V\gamma\alpha)\,S\beta\nabla_1 - Q_1(V\alpha\beta)\,S\,\gamma\nabla_1 = -Q_1\nabla_1\,S\,\alpha\beta\gamma,$$

whence putting $S\,\alpha\beta\gamma = -d\mathfrak{s}$ we get equation (9).

7. It will be observed that the above theorems have been proved only for cases where we can put $dQ = -Q_1 S\; d\rho\nabla_1$ i.e. when the space fluxes of Q are finite. If at any isolated point they are not finite this point must be shut off from the rest of the space by a small closed surface or curve as the case may be and this surface or curve must be reckoned as part of the boundary of the space. If at a surface (or curve) Q has a discontinuous value so that its derivatives are there infinite whereas on each side they are finite, this surface (or curve) must be considered as part of the boundary and each element of it will occur twice, i.e. once for the part of the space on each side.

In the case of the isolated points, if the surface integral or line integral round this added boundary vanish, we can of course cease to consider these points as singular. Suppose Q becomes infinite at the point $\rho = \alpha$. Draw a small sphere of radius a and also a sphere of unit radius with the point α for centre, and consider the small sphere to be the added boundary. Let $d\Sigma'$ be the element of the unit sphere cut off by the cone which has α for vertex and the element $d\Sigma$ of the small sphere for base. Then $d\Sigma = a^2 d\Sigma'$ and we get for the part of the surface integral considered $a^2 \iint Q_{d\Sigma}\, d\Sigma'$ where $Q_{d\Sigma}$ is the value of Q at the element $d\Sigma$. If then

$$\mathrm{Lt}_{T(\rho-\alpha)=0}\, T^2(\rho-\alpha) Q_{d\Sigma} U(\rho-\alpha) = 0$$

the point may be regarded as not singular. If the limiting expression is finite the added surface integral will be finite. If the expression is infinite the added surface integral will be generally but not always infinite. Similarly in the case of an added line integral if $\mathrm{Lt}_{T(\rho-\alpha)=0} T(\rho-\alpha) Q U(\rho-\alpha)$ is zero or finite, the added line integral will be zero or finite respectively (of course including in the term finite a possibility of zero value). If this expression be infinite, the added line integral will generally also be infinite.

This leads to the consideration of potentials which is given in § 9.

8. Some particular cases of equations (8) and (9) which (except the last) have been proved by Tait, are very useful. First put Q = a simple scalar P. Thus

$$\int P\, d\rho = \iint V\, d\Sigma\, \nabla P, \tag{10}$$

$$\iint P\, d\Sigma = \iiint \nabla P\, d\mathfrak{s}. \tag{11}$$

If P be the pressure in a fluid $-\iint P\, d\Sigma$ is the force resulting from the pressure on any portion and equation (11) shews that $-\nabla P$ is the force per unit volume due to the same cause. Next put $Q\omega = S\omega\sigma$ and $V\omega\sigma$. Thus

$$\int S\, d\rho\sigma = \iint S\, d\Sigma\, \nabla\sigma, \tag{12}$$

$$\int V\, d\rho\sigma = \iint V(V\, d\Sigma\, \nabla \,.\, \sigma) = \iint d\Sigma\, S\nabla\sigma - \iint \nabla_1 S\, d\Sigma\sigma_1, \tag{13}$$

$$\iint S\, d\Sigma\sigma = \iiint S\nabla\sigma\, d\mathfrak{s}, \tag{14}$$

$$\iint V\, d\Sigma\sigma = \iiint V\nabla\sigma\, d\mathfrak{s}. \tag{15}$$

Equations (12) and (14) are well-known theorems, and (13) and (15) will receive applications in the following pages. Green's Theorem with Thomson's extension of it are, as indeed has been pointed out by Tait particular cases of these equations.

Equations (14) and (15) applied to an element give the well-known physical meanings for $S\nabla\sigma$ and $V\nabla\sigma$*. The first is obtained by applying (14) to any element, and the second (regarding σ as a velocity) is obtained by applying (15) to

*[Note added, 1892. Let me disarm criticism by confessing that what follows concerning $V\nabla\sigma$ is nonsense.]

the element contained by the following six planes each passing infinitely near to the point considered—(1) two planes containing the instantaneous axis of rotation, (2) two planes at right angles to this axis, and (3) two planes at right angles to these four.

One very frequent application of equation (9) may be put in the following form:—Q being any linear function (varying from point to point) of R_1 and ∇_1, R being a function of the position of a point

$$\iiint Q(R_1, \nabla_1)\, d\mathfrak{s} = -\iiint Q_1(R, \nabla_1)\, d\mathfrak{s} + \iint Q(R, d\Sigma). \tag{16}$$

Potentials.

9. We proceed at once to the application of these theorems in integration to Potentials. Although the results about to be obtained are well-known ones in Cartesian Geometry or are easily deduced from such results it is well to give this quaternion method if only for the *collateral* considerations which on account of their many applications in what follows it is expedient to place in this preliminary section.

If R is some function of $\rho - \rho'$ where ρ' is the vector coordinate of some point under consideration and ρ the vector coordinate of any point in space, we have

$$_{\rho}\nabla R = -_{\rho'}\nabla R.$$

Now let $Q(R)$ be any function of R, the coordinates of Q being functions of ρ only. Consider the integral $\iiint Q(R)\, d\mathfrak{s}$ the variable of integration being ρ (ρ' being a constant so far as the integral is concerned). It does not matter whether the integral is a volume, surface or linear one but for conciseness let us take it as a volume integral. Thus we have

$$_{\rho'}\nabla \iiint Q(R)\, d\mathfrak{s} = \iiint {}_{\rho'}\nabla Q(R)\, d\mathfrak{s} = -\iiint {}_{\rho}\nabla_1 Q(R_1)\, d\mathfrak{s}.$$

Now $_{\rho}\nabla$ operating on the whole integral has no meaning so we may drop the affix to the ∇ outside and always understand ρ'. Under the integral sign however we must retain the affix ρ or ρ' unless a convention be adopted. It is convenient to adopt such a convention and since Q will probably contain some $_{\rho}\nabla$ but cannot possibly contain a $_{\rho'}\nabla$ we must assume that when ∇ appears without an affix under the integral sign the affix ρ is understood. With this understanding we see that when ∇ crosses the integral sign it must be made to change sign and refer only to the part we have called R. Thus

$$\nabla \iiint Q(R)\, d\mathfrak{s} = -\iiint \nabla_1 Q(R_1)\, d\mathfrak{s}. \tag{17}$$

Generally speaking R can and will be put as a function of $T^{-1}(\rho' - \rho)$ and for this we adopt the single symbol u. Both this symbol and the convention just explained will be constantly required in all the applications which follow.

10. Now let Q be any function of the position of a point. Then the potential of the volume distribution of Q, say q, is given by:—

$$q = \iiint uQ\, d\mathfrak{s}, \tag{18}$$

the extent of the volume included being supposed given. We may now prove the following two important propositions

$$\nabla^2 q = 4\pi Q, \tag{19}$$

$$\iint S\, d\Sigma\, \nabla \,.\, q = 4\pi \iiint Q\, d\mathfrak{s}. \tag{20}$$

The latter is a corollary of the former as is seen from equation (9) § 6 above.

Equation (19) may be proved thus. If P be any function of the position of a point which is finite but not necessarily continuous for all points

$$\nabla \iiint uP\, d\mathfrak{s},$$

is always finite and if the volume over which the integral extends is indefinitely diminished, so also is the expression now under consideration, and this for the point at which is this remnant of volume. This in itself is an important proposition. The expression, by equation (17), $= -\iiint \nabla uP\, d\mathfrak{s}$ and both statements are obviously true except for the point ρ'. For this point we have merely to shew that the part of the volume integral just given contributed by the volume indefinitely near to ρ' vanishes with this volume. Divide this near volume up into a series of elementary cones with ρ' for vertex. If r is the (small) height and $d\omega$ the solid vertical angle of one of these cones, the part contributed by this cone is approximately $U(\rho - \rho')P_{\rho'} r\, d\omega/3$ where $P_{\rho'}$ is the value of P at the point ρ'. The proposition is now obvious.

Now since

$$\nabla^2 q = \nabla^2 \iiint uQ\, d\mathfrak{s} = \iiint \nabla^2 uQ\, d\mathfrak{s},$$

we see that the only part of the volume integral $\iiint uQ\, d\mathfrak{s}$ which need be considered is that given by the volume in the immediate neighbourhood of ρ', for at all points except ρ', $\nabla^2 u = 0$. Consider then our volume and surface integrals only to refer to a small sphere with ρ' for centre and so small that no point is included at which Q is discontinuous and therefore ∇Q infinite. (This last assumes that Q

is not discontinuous at ρ'. In the case when Q is discontinuous at ρ' no definite meaning can be attached to the expression $\nabla^2 q$.) We now have

$$\begin{aligned}\nabla^2 q &= \nabla^2 \iiint uQ\, d\mathfrak{s} = -\nabla \iiint \nabla u\, Q\, d\mathfrak{s} \quad \text{[by § 9]}\\ &= \nabla \iiint u\, \nabla Q\, d\mathfrak{s} - \nabla \iint u\, d\Sigma\, Q,\end{aligned}$$

equation (9) § 6 being applied and the centre of the sphere being not considered as a singular point since the condition of § 7 is satisfied, viz. that $\mathrm{Lt}_{T(\rho-\rho')=0}\, T^2(\rho-\rho')uU(\rho-\rho') = 0$. Now putting P above $= \nabla Q$ we see that the first expression, viz. $\nabla \iiint u\, \nabla Q\, d\mathfrak{s}$ can be neglected and the second gives

$$\nabla^2 q = \iint \nabla u\, d\Sigma\, Q = 4\pi \overline{Q},$$

where $\overline{Q}$ is the mean value of Q over the surface of the sphere and therefore in the limit $= Q$. Thus equation (19) has been proved.

When Q has a simple scalar value all the above propositions, and indeed processes, become well-known ones in the theory of gravitational potential.

We do not propose to go further into the theory of Potentials as the work would not have so direct a bearing on what follows as these few considerations.

SECTION III.

*ELASTIC SOLIDS.

Brief recapitulation of previous work in this branch.

11. As far as I am aware the only author who has applied Quaternions to Elasticity is Prof. Tait. In the chapter on Kinematics of his treatise on Quaternions, §§ 360–371, he has considered the mathematics of strain with some elaboration and again in the chapter on Physical Applications, §§ 487–491, he has done the same with reference to stress and also its expression in terms of the displacement at every point of an elastic body.

In the former he has very successfully considered various useful decompositions of strain into pure and rotational parts and so far as strain alone is considered, i.e. without reference to what stress brings it about he has left little or nothing to be done. In the latter he has worked out the expressions for stress by means of certain vector functions at each point, which express the elastic properties of the body at that point.

But as far as I can see his method will not easily adapt itself to the solution of problems which have already been considered by other methods, or prepare the way for the solution of fresh problems. To put Quaternions in this position is our present object. I limit myself to the statical aspect of Elasticity, but I believe that Quaternions can be as readily, or nearly so, applied to the Kinetics of the subject.

For the sake of completeness I shall repeat in my own notation a small part of the work that Tait has given.

Tait shews (§ 370 of his *Quaternions*) that in any small portion of a strained medium the strain is homogeneous and (§ 360) that a homogeneous strain function is a linear vector one. He also shews (§ 487) that the stress function is a linear

*[Note added, 1892. It would be better to head this section "Elastic *bodies*" since except when the strains are assumed small the equations are equally true of solids and fluids. I may say here that I have proved in the *Proc. R. S. E.* 1890–91, pp. 106 *et seq.*, most of the general propositions of this section somewhat more neatly though the processes are essentially the same as here.]

vector one and he obtains expressions (§§ 487–8) for the force per unit volume due to the stress, in terms of the space-variation of the stress.

Strain, Stress-force, Stress-couple.

12. This last however I give in my own notation. His expression in § 370 for the strain function I shall throughout denote by χ. Thus

$$\chi\omega = \omega - S\,\omega\nabla\,.\,\eta, \tag{1}$$

where η is the displacement that gives rise to the strain.

Let χ consist of a pure strain ψ followed by a rotation $q(\ )q^{-1}$ as explained in Tait's *Quaternions*, § 365 where he obtains both q and ψ in terms of χ. Thus

$$\chi\omega = q\,\psi\omega\,q^{-1}. \tag{2}$$

When the strain is small ψ takes the convenient form $\overline{\chi}$ where χ stands for the pure part of χ so that

$$\overline{\chi}\omega = \omega - \tfrac{1}{2}S\,\omega\nabla\,.\,\eta - \tfrac{1}{2}\nabla_1 S\,\omega\eta_1, \tag{3}$$

by equation (3) § 2 above. Similarly $q(\ )q^{-1}$ becomes $V\theta(\ )$ where $2\theta = V\nabla\eta$. The truth of these statements is seen by putting in equation (2) for q, $1 + \theta/2$ and therefore for q^{-1}, $1 - \theta/2$ for ψ, $\overline{\chi}$ and then neglecting all small quantities of an order higher than the first.

13. Next let us find the force and couple per unit volume due to a stress which varies from point to point. Let the stress function be ϕ. Then the force on any part of the body, due to stress, is

$$\iint \phi\, d\Sigma = \iiint \phi\Delta\, d\mathfrak{s},$$

by equation (9) § 6. Thus *the force per unit volume* = $\phi\Delta$, for the volume considered in the equation may be taken as the element $d\mathfrak{s}$.

Again the moment round any arbitrary origin is

$$\iint V\rho\phi\, d\Sigma = \iiint V\rho_1\phi\nabla_1\, d\mathfrak{s} + \iiint V\rho\phi_1\nabla_1\, d\mathfrak{s},$$

by equation (9) § 6. The second term on the right is that due to the force $\phi\Delta$ just considered, and the first shews that in addition to this *there is a couple per unit volume* $= V\zeta\phi\zeta =$ *twice the "rotation" vector* of ϕ. Let then $\overline{\phi}$ be the pure part and $V\epsilon(\ )$ the rotatory part of ϕ. Thus

Force per unit volume

$$= \phi\Delta = \overline{\phi}\Delta - V\nabla\epsilon, \tag{4}$$

Couple per unit volume

$$= V\zeta\phi\zeta = 2\epsilon. \tag{5}$$

These results are of course equivalent to those obtained by Tait, *Quaternions*, §§ 487–8.

The meanings just given to $\eta, \chi, \overline{\chi}, \psi, q, \phi, \overline{\phi}$ and ϵ will be retained throughout this Section. In all cases of small strain as we have seen we may use ψ or $\overline{\chi}$ indifferently and whenever we wish to indicate that we are considering the physical phenomenon of pure strain we shall use ψ, $\overline{\chi}$ being regarded merely as a function of χ. We shall soon introduce a function ϖ which will stand towards ϕ somewhat as ψ towards χ and such that when the strain is small $\varpi = \overline{\phi}$.

It is to be observed that $\phi\omega$ is the force exerted on a vector area, which *when strained* is ω, not the stress on an area which before strain is ω. Similarly in equations (4) and (5) the independent variable of differentiation is $\rho + \eta$ so that strictly speaking in (4) we should put $\overline{\phi}_{\rho+\eta}\Delta - V_{\rho+\eta}\nabla\epsilon$. In the case of small strain these distinctions need not be made.

Stress in terms of strain.

14. To express stress in terms of strain we assume any displacement and consequent strain at every point of the body and then give to every point a small additional displacement $\delta\eta$ and find in terms of ψ and ϕ the increment $\iiint \delta w\, d\mathfrak{s}_0$ in the potential energy of the body, $w\, d\mathfrak{s}_0$ being the potential energy of any element of the body whose volume before strain was $d\mathfrak{s}_0$. Thus

$$\iiint \delta w\, d\mathfrak{s}_0 = \text{(work done by stresses on surface of portion considered)}$$
$$- \text{(work done by stresses throughout volume of same portion)}.$$

Thus, observing that by § 12 the rotation due to the small displacement $\delta\eta$ is $V_{\rho+\eta}\nabla\delta\eta/2$, we have

$$\iiint \delta w\, d\mathfrak{s}_0 = -\iint S\, \delta\eta\, \phi\, d\Sigma + \iiint S\, \delta\eta\, \phi_{1\rho+\eta}\nabla_1\, d\mathfrak{s} + \iiint S\, \epsilon_{\rho+\eta}\nabla\delta\eta\, d\mathfrak{s}.$$

The first of the terms on the right is the work done on the surface of the portion of the body considered; the second is $-$(work done by stress-forces $\phi_{\rho+\eta}\Delta$); and the third is −(work done by stress-couples 2ϵ). Thus converting the surface integral into a volume integral by equation (9) § 6 above

$$\begin{aligned}\iiint \delta w\, d\mathfrak{s}_0 &= -\iiint S\ \delta\eta_1\ \phi_{\rho+\eta}\nabla_1\, d\mathfrak{s} + \iiint S\ \delta\eta_1\ V\epsilon_{\rho+\eta}\nabla_1\, d\mathfrak{s}\\ &= -\iiint S\ \delta\eta_1\ \overline{\phi}_{\rho+\eta}\nabla_1\, d\mathfrak{s}.\end{aligned}$$

Limiting the portion of the body considered to the element $d\mathfrak{s}$ we get

$$\delta w = -mS\ \delta\eta_1\ \overline{\phi}_{\rho+\eta}\nabla_1 \tag{6}$$

where m is put for $d\mathfrak{s}/d\mathfrak{s}_0$ and therefore may be put by § $3a$ and § 3 above in the various forms

$$\begin{aligned}6m &= S\zeta_1\zeta_2\zeta_3\, S\chi\zeta_1\chi\zeta_2\chi\zeta_3 &&(6a)\\ &= S\zeta_1\zeta_2\zeta_3\, S\chi'\zeta_1\chi'\zeta_2\chi'\zeta_3 &&(6b)\\ &= S\zeta_1\zeta_2\zeta_3\, S\psi\zeta_1\psi\zeta_2\psi\zeta_3 &&(6c)\\ &= S\nabla_1\nabla_2\nabla_3\, S(\rho+\eta)_1(\rho+\eta)_2(\rho+\eta)_3. &&(6d)\end{aligned}$$

It is to be observed that since the rotation-vector ϵ of ϕ does not occur in equation (6), ϵ and therefore the stress-couple are quite arbitrary so far as the strain and potential energy are concerned. I do not know whether this has been pointed out before. Of course other data in the problem give the stress-couple. In fact it can be easily shewn that in all cases whether there be equilibrium or not the external couple per unit volume balances the stress-couple. [Otherwise the angular acceleration of the element would be infinite.] Thus if $\mathbf{M}$ be the external couple per unit volume of the *unstrained* solid we have always

$$\mathbf{M} + 2m\epsilon = 0. \tag{7}$$

In the particular case of equilibrium $\mathbf{F}$ being the external force per unit volume of the unstrained solid we have

$$\mathbf{F} + m(\overline{\phi}_{\rho+\eta}\Delta - V_{\rho+\eta}\nabla\epsilon) = 0. \tag{8}$$

The mathematical problem is then the same as if for $\mathbf{F}/m$ we substituted $\mathbf{F}/m + V_{\rho+\eta}\nabla(\mathbf{M}/2m)$ and for $\mathbf{M}$, zero. In the case of small strains m may be put $= 1$. In this case then the mathematical problem is the same as if for $\mathbf{F}$ we substituted $\mathbf{F} + V\nabla\mathbf{M}/2$ and for $\mathbf{M}$ zero.

Returning to equation (6) observe that

$$\delta\chi\omega = -S\chi\omega_{\rho+\eta}\nabla\,.\,\delta\eta$$

[which is established just as is the equation $\chi\omega = -S\,\omega\nabla\,.\,(\rho+\eta)$], so that changing ω which is *any* vector into $\chi^{-1}\omega$

$$\delta\chi\,.\,\chi^{-1}\omega = -S\,\omega_{\rho+\eta}\nabla\,.\,\delta\eta.$$

Therefore by equation (6) of this section and equation (4) § 3 above we have

$$\delta w = -mS\,\delta\chi\,\chi^{-1}\,\zeta\overline{\phi}\zeta.$$

We must express the differential on the right of this equation in terms of $\delta\psi$ and δq, the latter however disappearing as we should expect. Now $\chi\omega = q\psi\omega q^{-1}$ [equation (2) § 12] so that remembering that $\delta\,.\,q^{-1} = -q^{-1}\,\delta q\,q^{-1}$

$$\begin{aligned}\delta\chi\omega &= \delta q\,\psi\omega q^{-1} - q\psi\omega q^{-1}\,\delta q\,q^{-1} + q\,\delta\psi\,\omega q^{-1}\\ &= 2VV\,\delta q\,q^{-1}\,.\,\chi\omega + q\,\delta\psi\,\omega q^{-1},\end{aligned}$$

$$\therefore \qquad \delta w/m = -2S\,.\,V\,\delta q\,q^{-1}\,.\,\zeta\overline{\phi}\zeta - S\,q\,\delta\psi\,\chi^{-1}\zeta\,q^{-1}\overline{\phi}\zeta,$$

or since $V\,\zeta\overline{\phi}\zeta = 0$, $\overline{\phi}$ being self-conjugate

$$\delta w = -mS\;\delta\psi\,\chi^{-1}\,\zeta q^{-1}\overline{\phi}\,\zeta q.$$

This can be put into a more convenient shape for our present purpose. First put for χ^{-1} its value $\psi^{-1}q^{-1}(\;)q$ and then apply equation (6*a*) § 3 above, putting for the ϕ of that equation $q^{-1}(\;)q$ and therefore for the ϕ', $q(\;)q^{-1}$. Thus

$$\delta w = -mS\;\delta\psi\,\psi^{-1}\,\zeta\varpi\zeta, \tag{9}$$

where

$$\varpi\omega = q^{-1}\overline{\phi}(q\omega q^{-1})q. \tag{10}$$

The physical meaning of this last equation can easily be shewn. Suppose when there is no rotation that $\overline{\phi} = \varpi'$. Then it is natural* to assume—in fact it seems almost axiomatic—that the superimposed rotation $q(\;)q^{-1}$ should merely so to speak rotate the stress along with it. Thus if ω is some vector area before the rotation which becomes ω' by means of the rotation

$$q\varpi'\omega\,.\,q^{-1} = \overline{\phi}\omega'.$$

But $\omega' = q\omega q^{-1}$, so that

$$\varpi'\omega = q^{-1}\overline{\phi}(q\omega q^{-1})q = \varpi\omega.$$

*Observe that we *do not* make this assumption. We really shew that it is true.

Thus we see that ϖ is as it were $\bar{\phi}$ with the rotation undone.

Before proceeding further with the calculation let us see what we have assumed and what equation (9) teaches us. The one thing we have assumed is that the potential energy of the body can be taken as the sum of the potential energies of its elements, in other words that no part of the potential energy depends conjointly on the strains at P and Q where P and Q are points separated by a finite distance. This we must take as an axiom. By it we are led to the expression for δw in equation (9). This only involves the variation of the pure strain ψ *but not the space differential coefficients of* ψ. This is not an obvious result as far as I can see but it is I believe always assumed without proof.

We may *now* regard w as a function of ψ only. Therefore by equation (7) § 5 above

$$\delta w = -S\ \delta\psi\,\zeta\,{}_{\psi}\text{ᗡ}w\,\zeta,$$

therefore by equation (9) of this section

$$S\ \delta\psi\,\zeta\,{}_{\psi}\text{ᗡ}w\,\zeta = mS\ \delta\psi\,\psi^{-1}\,\zeta\varpi\zeta^{*}.$$

In equation (6*a*) § 3 above putting $\phi = \psi^{-1}$ the right-hand member of this equation becomes $mS\ \delta\psi\,\zeta\,\varpi\psi^{-1}\,\zeta$. Now putting in equation (6*a*) § 3 $\phi = \varpi\psi^{-1}$ this member becomes $mS\,\zeta\,\delta\psi\,\psi^{-1}\,\varpi\zeta$ or $mS\ \delta\psi\,\zeta\,\psi^{-1}\varpi\,\zeta$. Thus we have

$$2S\ \delta\psi\,\zeta\,{}_{\psi}\text{ᗡ}w\,\zeta = mS\ \delta\psi\,\zeta(\varpi\psi^{-1} + \psi^{-1}\varpi)\zeta.$$

Now $\varpi\psi^{-1} + \psi^{-1}\varpi$ is self-conjugate. Hence by § 4 above

$$m(\varpi\psi^{-1} + \psi^{-1}\varpi) = 2{}_{\psi}\text{ᗡ}w. \tag{11}$$

This equation can be looked upon as giving ϖ in terms of the strain. We can obtain ϖ however explicitly for $\psi^{-1}\varpi$ is the conjugate of $\varpi\psi^{-1}$. Hence from equation (11) (because $(\psi^{-1}\varpi + \varpi\psi^{-1})/2$ is the pure part of $\varpi\psi^{-1}$)

$$m\,\varpi\psi^{-1}\omega = {}_{\psi}\text{ᗡ}w\,\omega + V\theta\,\omega$$

where θ is a vector to be found. Changing ω into $\psi\omega$

$$m\,\varpi\omega = {}_{\psi}\text{ᗡ}w\,\psi\omega + V\theta\,\psi\omega.$$

*By putting $\delta\psi = \omega S(\ )\omega' + \omega' S(\ )\omega$ in this equation, equation (11) can be deduced but as this method has already been applied in § 4 I give the one in the text to shew the variety of Quaternion methods. [Note added, 1892. If we use the theorem in the foot-note of § 4, equation (11) follows at once.]

Now ϖ being self-conjugate $V\zeta\varpi\zeta = 0$. Hence

$$V\zeta{}_{\psi}\mathrm{ɑw}\,\psi\zeta = -V\zeta\,V\theta\,\psi\zeta = \theta S\,\zeta\psi\zeta - \psi\zeta\,S\,\theta\zeta$$
$$= \theta S\,\zeta\psi\zeta + \psi\theta,$$

whence

$$\left.\begin{aligned} &\theta = (\psi + S\,\zeta_1\psi\zeta_1)^{-1}V\zeta{}_{\psi}\mathrm{ɑw}\,\psi\zeta \\ \text{or}\qquad &\theta = (\psi + S\,\zeta_1\psi\zeta_1)^{-1}V\psi\zeta{}_{\psi}\mathrm{ɑw}\,\zeta \end{aligned}\right\} \tag{12}$$

by equation (6a) § 3 above. Thus finally

$$\left.\begin{aligned} &m\varpi\omega = {}_{\psi}\mathrm{ɑw}\,\psi\omega + V\theta\psi\omega \\ \text{or}\qquad &m\varpi\omega = {}_{\psi}\mathrm{ɑw}\,\psi\omega - V\psi\omega(\psi + S\,\zeta_1\psi\zeta_1)^{-1}V\psi\zeta{}_{\psi}\mathrm{ɑw}\,\zeta. \end{aligned}\right\} \tag{13}$$

This completely solves the problem of expressing stress in terms of strain in the most general case.

15. $\varpi\omega + V\epsilon'\omega$, where, as we saw in last section, ϵ' is perfectly arbitrary so far as the strain is concerned, is the force on the strained area ω due to the pure strain ψ. And again

$$\bar{\phi}\omega + V\epsilon''\omega \quad \text{or} \quad q(\varpi q^{-1}\,\varpi q)q^{-1} + V\epsilon''\omega$$

is that due to the strain $q\psi(\;)q^{-1}$ or χ.

To find the force on an area which before strain was ω_0 let its strained value after ψ has taken place be ω. Then by equation (4), § 145 of Tait's *Quaternions*, $m\omega_0 = \psi\omega$. Hence

$$\text{Required force} = \varpi\omega + V\epsilon'\omega = {}_{\psi}\mathrm{ɑw}\,\omega_0 + V\theta\omega_0 + mV\epsilon'\psi^{-1}\omega_0$$

by equation (13) of last section. If the rotation now take place this force rotates with it so that the force on the area which was originally ω_0 is after the strain χ or $q\psi(\;)q^{-1}$

$$\tau\omega_0 = q{}_{\psi}\mathrm{ɑw}\,\omega_0 q^{-1} + qV\theta\omega_0\,.\,q^{-1} + V(\epsilon q\psi^{-1}\omega_0\,.\,q^{-1}) \tag{14}$$

where $\epsilon = m\,q\epsilon' q^{-1}$ and

$\therefore \qquad \epsilon = \frac{1}{2}$ the stress-couple per un. vol. of unstrained body.

This force then is a linear vector function of ω_0, but in general even when $\epsilon = 0$ it is not self-conjugate. When both ϵ and the rotation are zero we see that the rotation vector of τ is θ given by equation (12) of last section.

The stress-force can be shewn as in § 13 to be $\tau\Delta$ per unit volume of the unstrained body. Thus since the corresponding stress-couple is 2ϵ the moment exerted by the stresses on any portion of the body round an arbitrary origin is

$$\iiint V(\rho + \eta)\tau_1\nabla_1\, d\mathfrak{s}_0 + 2\iiint \epsilon\, d\mathfrak{s}_0.$$

But this moment may also be put in the form

$$\iint V(\rho + \eta)\tau\, d\Sigma_0,$$

or $$\iiint V\zeta\tau\zeta\, d\mathfrak{s}_0 + \iiint V(\rho + \eta)\tau_1\nabla_1\, d\mathfrak{s}_0 + \iiint V\eta_1\tau\nabla_1\, d\mathfrak{s}_0,$$

by equation (9) § 6 above. Comparing these results

$$\left.\begin{aligned} 2\epsilon &= V\zeta\tau\zeta + V\eta_1\tau\nabla_1 \\ \text{or}\quad 2\epsilon &= V\rho_1'\tau\nabla_1, \end{aligned}\right\} \tag{14a}$$

in the notation of next section. This equation may also be deduced from equation (15l) below but not so naturally as above.

The equations of equilibrium.

16. We require equation (9) § 14 above to prove the statement that no space-variations of ψ or q are involved in w. It is also required to shew that the fact that q also is not involved in w is a mathematical sequence of the assumption that the potential energy of a solid is the sum of the potential energies of its elements. Assuming these facts however we can arrive at the equation of stress (11) § 14 in a different way from the above. We shall also obtain quite different expressions for ϖ, τ, &c. and most important of all we shall obtain the equations of equilibrium by obtaining τ explicitly in terms of the displacement and its space derivatives. From § 12 above we have

$$\chi\omega = q\psi\omega q^{-1}, \qquad \chi'\omega = \psi(q^{-1}\omega q),$$
$$\therefore \qquad \chi'\chi = \psi^2 = \Psi, \tag{15}$$

(say) as in Tait's *Quaternions*, § 365. Thus $\Psi\omega = \nabla_1 S\omega\nabla_2 S\rho_1'\rho_2'$ where ρ' is put as it will be throughout this section for $\rho + \eta$, the vector coordinate after displacement of the point ρ. From this as we have seen in the Introduction we deduce that the coordinates of Ψ are the A, B, C, a, b, c of Thomson and Tait's *Nat. Phil.* App. C.

We may as do those authors regard w as a function of Ψ. Thus:—

$$\delta w = -S\,\delta\Psi\,\zeta_{\Psi}\text{ᗡ}w\,\zeta,$$
$$= -S\,\delta\chi'\chi\zeta_{\Psi}\text{ᗡ}w\,\zeta - S\chi'\,\delta\chi\,\zeta_{\Psi}\text{ᗡ}w\,\zeta.$$

By equation (4a) § 3 above, each of the terms in this last expression

$$= -S\,\delta\eta_1\chi_{\Psi}\text{ᗡ}w\,\nabla_1,$$

$$\therefore \qquad \delta w = -2S\,\delta\eta_1\chi_{\Psi}\text{ᗡ}w\,\nabla_1, \tag{15a}$$

Comparing this with equation (6) § 14 and putting in both $\delta\eta_1 = \omega' S\,\omega\rho'$ where ω', ω are arbitrary constant vectors, we get

$$mS\,\omega'\,\overline{\phi}\omega = 2S\,\omega'\chi_{\Psi}\text{ᗡ}w\,\chi'\omega.$$

Hence, since ω' is quite arbitrary

$$m\overline{\phi} = 2\chi_{\Psi}\text{ᗡ}w\,\chi'. \tag{15b}$$

From equation (10) § 14 above which *defines* ϖ we see that

$$m\overline{\varpi} = 2\psi_{\Psi}\text{ᗡ}w\,\psi, \tag{15c}$$

which we should expect since we have already seen that ϖ is the value of $\overline{\phi}$ when there is no rotation and therefore $\chi = \chi' = \psi$. Now since

$$S\,\delta\psi\,\zeta_{\psi}\text{ᗡ}\zeta = S\,\delta\Psi\,\zeta_{\Psi}\text{ᗡ}\zeta = S\,(\delta\psi\,\psi + \psi\,\delta\psi)\zeta_{\Psi}\text{ᗡ}\zeta,$$

we deduce by any one of the processes already exemplified that

$$_{\psi}\text{ᗡ} = {}_{\Psi}\text{ᗡ}\,\psi + \psi_{\Psi}\text{ᗡ},$$

where of course the differentiations of $_{\Psi}\text{ᗡ}$ must not refer to ψ. We see then from equation (15c) that

$$m(\varpi\psi^{-1} + \psi^{-1}\varpi) = 2_{\psi}\text{ᗡ}w,$$

which is equation (11) § 14.

Our present purpose however is to find the equations of equilibrium. Let ω_0 be the vector area which by the strain χ becomes ω. Thus* as in last section

$$m\omega_0 = \chi'\omega. \tag{15e}$$

*See § 83 below.

Further let 2ϵ be the stress-couple per unit volume of the unstrained solid so that $2\epsilon/m$ is the same of the strained solid. As we know, ϵ is quite independent of the strain. By § 13 we see that the force $\tau\omega_0$ on the area which before strain was ω_0 is $\overline{\phi}\omega + V\epsilon\omega/m$. Therefore

$$\tau\omega_0 = 2\chi{}_{\Psi}\sqsupset w\,\omega_0 + V\epsilon\chi'^{-1}\omega_0. \tag{15f}$$

We saw in the last section that the force per unit volume of the unstrained solid is $\tau\Delta$ and the couple 2ϵ. Hence

$$\mathbf{F} + \tau\Delta = 0, \tag{15g}$$

$$\mathbf{M} + 2\epsilon = 0, \tag{15h}$$

are the equations of equilibrium, where **F** and **M** are the external force and couple per unit volume of the unstrained solid. All that remains to be done then is to express τ in terms of ϵ and the displacement. Putting

$$\rho + \eta = \rho', \tag{15i}$$

as already mentioned, we have

$$\chi\omega = -\rho'_1 S\,\omega\nabla_1, \tag{15j}$$

so that by equation (6*h*) § 3*a* above we have

$$\chi'^{-1}\omega = -3V\rho'_1\rho'_2\, S\,\omega\nabla_1\nabla_2/S\,\nabla_1\nabla_2\nabla_3\, S\rho'_1\rho'_2\rho'_3. \tag{15k}$$

Therefore by equation (15*f*)

$$\tau\omega = -2\rho'_1 S\,\nabla_1{}_{\Psi}\sqsupset w\,\omega - 3V\epsilon\, V\rho'_1\rho'_2\, S\,\omega\nabla_1\nabla_2/S\,\nabla_1\nabla_2\nabla_3\, S\rho'_1\rho'_2\rho'_3. \tag{15l}$$

It is unnecessary to write down what equation (15*g*) becomes when we substitute for τ, changing ϵ into $-\mathbf{M}/2$ by equation (15*h*). In the important case however when $\mathbf{M} = 0$, the equation is quite simple, viz.

$$\mathbf{F} = 2\rho'_1 S\,\nabla_1{}_{\Psi}\sqsupset w\,\Delta. \tag{15m}$$

Addition to § 16, Dec., 1887 (sent in with the Essay). [The following considerations occurred just before I was obliged to send the essay in, so that though I thought them worth giving I had not time to incorporate them in the text.

It is interesting to consider the case of an isotropic body. Here w is a function of the three principal elongations only and therefore we may in accordance with

§ 14 and § 15 suppose it a function of a, b, c or in accordance with § 16 of A, B, C where

$$a = -S\zeta\psi\zeta, \qquad b = -S\zeta\psi^2\zeta/2, \qquad c = -S\zeta\psi^3\zeta/3. \tag{A}$$

$$A = -S\zeta\Psi\zeta, \qquad B = -S\zeta\Psi^2\zeta/2, \qquad C = -S\zeta\Psi^3\zeta/3. \tag{B}$$

Let us use x, y, z, X, Y, Z for the differential coefficients of w with respect to a, b, c, A, B, C respectively. Thus

$$dw = x\,da + y\,db + z\,dc = -S\,d\psi\,\zeta(x + y\psi + z\psi^2)\zeta$$

as can be easily proved by means of equation (6a) § 3 above. But

$$dw = -S\,d\psi\zeta_{\psi}\daleth w\,\zeta.$$

$$\therefore \qquad {}_{\psi}\daleth w = x + y\psi + z\psi^2, \tag{C}$$

by § 4 above. Similarly we have

$${}_{\Psi}\daleth w = X + Y\Psi + Z\Psi^2. \tag{D}$$

(Notice in passing that to pass to small strains is quite easy for ${}_{\psi}\daleth w$ is linear and homogeneous in $\psi - 1$ so that ${}_{\psi}\daleth w = x + y\psi$ where y is a constant and $x + y$ a multiple of $a - 3$.)*

From equation (C) we see that θ in equation (12) § 14 is zero and therefore that from equation (13)

$$m\varpi = {}_{\psi}\daleth w\,\psi = x\psi + y\psi^2 + z\psi^3. \tag{E}$$

Again from equation (14) § 15

$$\tau\omega = xq\omega q^{-1} + y\chi\omega + z\chi\psi\omega + V\epsilon\chi'^{-1}\omega. \tag{F}$$

Similarly from equations (15b) and (15f) § 16 we may prove by equation (D) that

$$\tfrac{1}{2}m\bar{\phi} = \mathbf{X}\chi\chi' + \mathbf{Y}(\chi\chi')^2 + \mathbf{Z}(\chi\chi')^3, \tag{G}$$

and

$$\tau = 2X\chi + 2Y\chi\chi'\chi + 2Z\chi\chi'\chi\chi'\chi + V\epsilon\chi'^{-1}(\;). \tag{H}$$

If we wish to neglect all small quantities above a certain order the present equations pave the way for suitably treating the subject. I do not however propose to consider the problem here as I have not considered it sufficiently to do it justice.]

---

*[Note added, 1892. In the original essay there was a slip here which I have corrected. It was caused by assuming that ψ instead of $\psi - 1$ was small for small strains. In the original I said "where y is a constant and x is a multiple of a."]

Variation of Temperature.

16a. The w which appears in the above sections is the same as the w which occurs in Tait's *Thermo-dynamics*, § 209, and therefore all the above work is true whether the solid experience change of temperature or not. w will be a function then of the temperature as well as ψ. To express the complete mathematical problem of the physical behaviour of a solid we ought of course instead of the above equations of equilibrium, to have corresponding equations of motion, viz. equations (15h), (15l) and (instead of (15g))

$$D\ddot{\rho}' = \mathbf{F} + \tau\Delta, \tag{15n}$$

where D is the original density of the solid at the point considered. Further we ought to put down the equations of conduction of heat and lastly equations (16f) and (16d) below.

We do not propose to consider the conduction of heat, but it will be well to shew how the thermo-dynamics of the present question are treated by Quaternions.

Let t be the temperature of the element which was originally $d\varsigma_0$ and $E\,d\varsigma_0$ its intrinsic energy. Let

$$\delta H = -S\,\delta\psi\,\zeta M\zeta + N\,\delta t,$$

where M, N are a linear self-conjugate function and a scalar respectively, both functions of ψ and t. Here $\delta H\,d\varsigma_0$ is the heat required to be put into the element to raise its temperature by δt and its pure strain by $\delta\psi$. Now when t is constant δH must be a perfect differential so that we may put

$$M = t_\psi \alpha f,$$

where f is some function of ψ and t. Thus

$$\delta H = -tS\,\delta\psi\,\zeta_\psi \alpha f\,\zeta + N\,\delta t. \tag{16}$$

Now we have seen in § 14 that the work done on the element during the increment $\delta\psi$, divided by $d\varsigma_0$

$$= -mS\,\delta\psi\,\psi^{-1}\,\zeta\varpi\zeta.$$

Thus by the first law of Thermo-dynamics

$$\begin{aligned} \delta E &= J\,\delta H - mS\,\delta\psi\,\psi^{-1}\,\zeta\varpi\zeta \\ &= JN\,\delta t - JtS\,\delta\psi\,\zeta_\psi \alpha f\,\zeta - mS\,\delta\psi\,\psi^{-1}\,\zeta\varpi\zeta, \end{aligned}$$

where J is Joule's mechanical equivalent. Thus

$$JN = dE/dt, \tag{16a}$$

and
$$\tfrac{1}{2}m(\varpi\psi^{-1} + \psi^{-1}\varpi) = {}_{\psi}\text{Ɑ}w, \tag{16b}$$

where
$$w = E - Jtf. \tag{16c}$$

To apply the second law we go through exactly the same cycle as does Tait in his *Thermo-dynamics*, § 209, viz.

$$(\psi, t)(\psi + \delta\psi, t)(\psi + \delta\psi, t + \delta t)(\psi, t + \delta t)(\psi, t).$$

We thus get*

$$J_{\psi}\text{Ɑ}f = -\tfrac{1}{2}m\left(\frac{d\varpi}{dt}\psi^{-1} + \psi^{-1}\frac{d\varpi}{dt}\right) = -\frac{d_{\psi}\text{Ɑ}w}{dt} = -{}_{\psi}\text{Ɑ}\frac{dw}{dt}$$

or
$$Jf = -dw/dt \tag{16d}$$

the arbitrary function of t being neglected as not affecting any physical phenomenon. Substituting for w from equation (16c),

$$Jt\,df/dt = dE/dt = JN \tag{16e}$$

by equation (16a). Thus from equation (16)

$$\delta H = t\,\delta f, \tag{16f}$$

so that in elastic solids as in gases we have a convenient function which is called the "entropy". Thus the intrinsic energy $E\,d\mathfrak{s}_0$, the entropy $f\,d\mathfrak{s}_0$ and the stress ϖ have all been determined in terms of one function w of ψ and t which function is therefore in this general mathematical theory supposed to be known.

If instead of regarding w (which with the generalised meaning it now bears may still conveniently be called the "potential energy" per unit volume) as the fundamental function of the substance we regard the intrinsic energy or the entropy as such it will be seen that one other function of ψ must also be known. For suppose f the entropy be regarded as known. Then since $dw/dt = -Jf$

$$w = W - J\textstyle\int f\,dt, \tag{16g}$$

*[By assuming from the second law that the work done *by* the element in the cycle, i.e. the sum of the works done by it during the first and third steps is $J\delta t/t$ multiplied by the heat absorbed by the element in the third step. Note added, 1893.]

where the integral is any *particular* one and W is a function of ψ only, supposed known. Again

$$E = w + Jtf$$

or

$$E = W + J(tf - \textstyle\int f\,dt). \tag{16h}$$

Thus all the functions are given in terms of f and W. Similarly if E be taken as the fundamental function

$$w = t(W' - \textstyle\int E\,dt/t^2), \tag{16i}$$

$$Jf = E/t + \textstyle\int E\,dt/t^2 - W', \tag{16j}$$

where as before the integral is some particular one and W' is a function of ψ only.

Small strains.

17. We now make the usual assumption that the strains are so small that their coordinates can be neglected in comparison with ordinary quantities such as the coefficients of the linear vector function ${}_{\psi}\text{Ɑ}w$. We can deduce this case from the above more general results.

To the order considered $q = 1$ so that by equation (10) § 14 above

$$\bar{\phi} = \varpi.$$

We shall use the symbol ϖ rather than $\bar{\phi}$ for the same reasons explained in § 13 above as induce us to use ψ rather than $\bar{\chi}$.

Remembering that ω_0 and ω are now the same and that ψ may be put $= 1$ so that $V\zeta_{\psi}\text{Ɑ}w\,\psi\zeta = V\zeta_{\psi}\text{Ɑ}w\,\zeta = 0$ and therefore θ of equation (12) $= 0$, we have from equation (13),

$$\varpi\omega = {}_{\psi}\text{Ɑ}w\,\omega. \tag{17}$$

Of course we do not require to go through the somewhat complicated process of § 14, § 15 to arrive at this result. In fact in equation (6) § 14, we may put $m = 1$ and ${}_{\rho+\eta}\nabla = \nabla$ so that

$$\delta w = -S\,\delta\eta_1\,\bar{\phi}\nabla_1 = -S\,\delta\eta_1\,\varpi\nabla_1,$$

and therefore by § 3 above

$$\delta w = -S\,\delta\bar{\chi}\,\zeta\varpi\zeta = -S\,\delta\psi\,\zeta\varpi\zeta.$$

But by § 5 above

$$\delta w = -S\,\delta\psi\,\zeta_{\psi}\text{Ɑ}w\,\zeta,$$

and therefore by § 4 we get equation (17).

18. It is convenient here to slightly change the notation. For ψ, χ we shall now substitute $\psi + 1$, $\chi + 1$ respectively. This leads to no confusion as will be seen.

With this notation the strain being small the stress is linear in ψ i.e. ${}_{\psi}\daleth w$ is linear and therefore w quadratic. Now for any such quadratic function

$$w = -S\psi\zeta_{\psi}\daleth w\zeta/2, \tag{18}$$

for we have by § 5,

$$dw = -S\,d\psi\,\zeta_{\psi}\daleth w\,\zeta.$$

Put now $\psi = n\psi'$. Then because ${}_{\psi}\daleth w$ is linear in ψ

$${}_{\psi}\daleth w = n_{\psi}\daleth' w,$$

where ${}_{\psi}\daleth' w$ is put for the value of ${}_{\psi}\daleth w$ when the coordinates of ψ' are substituted for those of ψ. Thus keeping ψ' constant and varying n,

$$dw = -n\,dn\,S\psi'\,\zeta_{\psi}\daleth' w\,\zeta,$$

whence integrating from $n = 0$ to $n = 1$ and changing ψ' into ψ we get equation (18).

From equation (18) we see that for small strains we have

$$w = -S\psi\zeta\varpi\zeta/2. \tag{19}$$

Now w is quadratic in ψ and therefore also quadratic in ϖ, so that regarding w as a function of ϖ we have as in equation (18)

$$w = -S\varpi\zeta_{\varpi}\daleth w\,\zeta/2,$$

so that by equation (19) and § 4 above

$$\psi' = {}_{\varpi}\daleth w. \tag{20}$$

All these results for small strains are well-known in their Cartesian form, but it cannot be bias that makes these quaternion proofs appear so much more natural and therefore more simple and beautiful than the ordinary ones.

19. Let us now consider (as in § 16 is really done) w as a function of the displacement. Now w is quadratic in ψ, and ψ is linear and symmetrical in ∇_1 and η_1. In fact from equation (3) § 12 above, remembering that the $\overline{\chi}$ of that equation is our present $\psi + 1$, we have

$$2\psi\omega = -\eta_1 S\omega\nabla_1 - \nabla_1 S\omega\eta_1.$$

Therefore we may put

$$w = w(\eta_1, \nabla_1, \eta_2, \nabla_2), \tag{21}$$

where $w(\alpha, \beta, \gamma, \delta)$ is linear in each of its constituents, is symmetrical in α and β, and again in γ and δ, and is also such that the pair α, β and the pair γ, δ can be interchanged. [This last statement can be *made* true if not so at first, by substituting for $w(\alpha, \beta, \gamma, \delta)$, $w(\alpha, \beta, \gamma, \delta)/2 + w(\gamma, \delta, \alpha, \beta)/2$ as this does not affect equation (21).] Such a function can be proved to involve 21 independent scalars, which is the number also required to determine an arbitrary quadratic function of ψ, since ψ involves six scalars.

Thus we have the two following expressions for δw, which we equate

$$\begin{aligned} -S\,\delta\psi\,\zeta\varpi\zeta &= w(\delta\eta_1, \nabla_1, \eta_2, \nabla_2) + w(\eta_1, \nabla_1, \delta\eta_2, \nabla_2) \\ &= 2w(\delta\eta_1, \nabla_1, \eta_2, \nabla_2), \end{aligned}$$

or* by § 3 above,

$$-S\,\delta\eta_1\,\varpi\nabla_1 = 2w(\delta\eta_1, \nabla_1, \eta_2, \nabla_2).$$

Now let us put $\delta\eta = \omega' S\omega\rho$ where ω' and ω are arbitrary constant vectors. We thus get

$$S\omega'\,\varpi\omega = -2w(\omega', \omega, \eta_1, \nabla_1) = 2S\omega'\,\zeta w(\zeta, \omega, \eta_1, \nabla_1).$$

Whence since ω' is quite arbitrary,

$$\varpi\omega = 2\zeta w(\zeta, \omega, \eta_1, \nabla_1). \tag{22}$$

The statical problem can now be easily expressed. As we saw in § 14, equations (7) and (8), it is simply

$$\mathbf{F} + V\nabla\mathbf{M}/2 + \varpi\Delta = 0, \tag{22a}$$

*[Note added, 1892. Better thus:—by § 3 above,

$$-S\,\delta\psi\,\zeta\varpi\zeta = 2w(\delta\psi\,\zeta, \zeta, \eta_1, \nabla_1) = -2S\zeta_1\,\delta\psi\,\zeta w(\zeta_1, \zeta, \eta_1, \nabla_1)$$

therefore by § 4, $\varpi\omega = 2\zeta w(\zeta, \omega, \eta_1, \nabla_1)$

for $\zeta w(\zeta, \omega, \eta_1, \nabla_1)$ regarded as a function of ω is clearly self conjugate.]

throughout the mass; and at the surface

$$\mathbf{F}_S - VU\,d\Sigma\,\mathbf{M}/2 - \varpi U\,d\Sigma = 0, \tag{22b}$$

where **F**, **M** are the given external force and couple per unit volume and $\mathbf{F}_S$ is the given external surface traction per unit surface. Substituting for ϖ from equation (22)

$$\left.\begin{aligned} \mathbf{F} + V\nabla\mathbf{M}/2 + 2\zeta w(\zeta, \Delta, \eta_1, \nabla_1) = 0 \\ \mathbf{F}_S - VU\,d\Sigma\,\mathbf{M}/2 - 2\zeta w(\zeta, U\,d\Sigma, \eta_1, \nabla_1) = 0. \end{aligned}\right\} \tag{23}$$

Isotropic Bodies.

20. The simplest way to treat these bodies is to consider the (linear) relations between ϖ and ψ.

In the first place notice that ψ can always be decomposed into three real elongations (contractions being of course considered as negative elongations). Thus i being the unit vector in the direction of such an elongation,

$$\psi\omega = -\Sigma\, ei\, S\, i\, \omega.$$

The elongation $-ei\,S\,i\,\omega$ will cause a stress symmetrical about the vector i, i.e. a tension Ae in the direction of i and a pressure Be in all directions at right angles; A and B being constants (on account of the linear relation between ϖ and ψ) independent of the direction of i (on account of the isotropy of the solid). This stress may otherwise be described as a tension $(A + B)e$ in the direction of i and a hydrostatic pressure Be. Thus

$$\begin{aligned} \varpi\omega &= -(A + B)\Sigma\, ei\, S\, i\, \omega - B\omega\,\Sigma e \\ &= (A + B)\psi\omega + B\omega\, S\,\zeta\,\psi\zeta. \end{aligned}$$

To obtain the values of A and B in terms of Thomson and Tait's coefficients k and n of cubical expansion and rigidity respectively; first put

$$\psi\omega = e\omega \quad \text{and} \quad \varpi\omega = 3k\,e\omega,$$

and then put

$$\psi\omega = V\lambda\omega\mu \quad \text{and} \quad \varpi\omega = 2nV\lambda\omega\mu,$$

λ and μ being any two vectors perpendicular to each other. We thus get

$$A + B = 2n, \qquad B = -(k - 2n/3),$$

whence

$$\varpi\omega = 2n\,\psi\omega - (k - 2n/3)\omega S\,\zeta\psi\zeta. \tag{24}$$

From this we have

$$\psi\omega = \varpi\omega/2n + \omega S\,\zeta\psi\zeta(k - 2n/3)/2n,$$

but from the same equation

$$S\,\zeta\varpi\zeta = 3k\,S\,\zeta\psi\zeta;$$

$$\therefore \qquad \psi\omega = \frac{1}{2n}\varpi\omega + \left(\frac{1}{6n} - \frac{1}{9k}\right)\omega S\,\zeta\varpi\zeta. \tag{25}$$

Equation (24) gives stress in terms of strain and (25) the converse.

21. We can now give the various useful forms of w for isotropic bodies for from equation (19) § 18,

$$w = -S\,\psi\,\zeta\varpi\zeta/2.$$

Therefore from equations (24) and (25) respectively

$$w = -n(\psi\zeta)^2 + \tfrac{1}{2}(k - 2n/3)S\,\zeta_1\psi\zeta_1\,S\,\zeta_2\psi\zeta_2, \tag{26}$$

$$w = -\frac{1}{4n}(\varpi\zeta)^2 - \tfrac{1}{2}\left(\frac{1}{6n} - \frac{1}{9k}\right)S\,\zeta_1\varpi\zeta_1\,S\,\zeta_2\varpi\zeta_2. \tag{27}$$

Therefore again from § 3 above and from equation (26),

$$w = -nS\,\eta_1\psi\nabla_1 + \tfrac{1}{2}(k - 2n/3)(S\,\nabla\eta)^2$$

or

$$\begin{aligned} 2w(\eta_1, \nabla_1, \eta_2, \nabla_2) = nS\,\nabla_1\eta_2\,S\,\nabla_2\eta_1 + nS\,\nabla_1\nabla_2\,S\,\eta_1\eta_2 \\ +(k - 2n/3)S\,\nabla_1\eta_1 S\,\nabla_2\eta_2. \end{aligned} \tag{28}$$

Hence from equation (22)

$$\varpi\omega = -nS\,\omega\nabla\,.\,\eta - n\nabla_1 S\,\eta_1\omega - (m - n)\omega S\,\nabla\eta, \tag{29}$$

where m is put for $k + n/3$. This last could have been deduced at once from equation (24) by substituting for $\psi\omega$.

Thus the equations (23) for the statical problem are

$$\mathbf{F} + V\nabla\mathbf{M}/2 = n\nabla^2\eta + m\nabla S\,\nabla\eta, \tag{30}$$

$$-\mathbf{F}_S + VU\,d\Sigma\,\mathbf{M}/2 = nS\,U\,d\Sigma\,\nabla\,.\,\eta + n\nabla_1 S\eta_1 U\,d\Sigma + (m-n)U\,d\Sigma\,S\,\nabla\eta. \tag{31}$$

We now proceed to apply these results for small strains in isotropic bodies to particular cases. These particular cases have all been worked out by the aid of Cartesian Geometry and they are given to illustrate the truth of the assertion made in the Introduction that the consideration of *general problems* is made simpler by the use of Quaternions instead of the ordinary methods.

Particular integral of equation (30)*.

22. Since from equation (30) (**F** being put for simplicity instead of **F** + $V\nabla\mathbf{M}/2$) we have

$$n\nabla^2\eta = \mathbf{F} - m\nabla S\,\nabla\eta$$

we obtain as a particular case by equations (18) and (19) § 10,

$$4\pi n\eta = \iiint u(\mathbf{F} - m\nabla S\,\nabla\eta)\,d\varsigma,$$

where u has the meaning explained in § 9, and the volume integral extends over any portion (say the whole) of the body we may choose to consider. To express $\iiint u\nabla S\,\nabla\eta\,d\varsigma$ as a function of **F** put in this term $u = -\frac{1}{2}U\rho_1\nabla_1$ where ρ is taken for the $\rho - \rho'$ of § 9, and apply equation (9) § 6. Thus

$$\begin{aligned}4\pi n\eta &= \iiint u\mathbf{F}\,d\varsigma + \tfrac{1}{2}m\iiint U\rho_1\nabla_1\nabla S\,\nabla\eta\,d\varsigma\\ &= \iiint u\mathbf{F}\,d\varsigma - \tfrac{1}{2}m\iiint U\rho\nabla^2 S\,\nabla\eta\,d\varsigma + \text{ a surf. int.}\\ &= \iiint u\mathbf{F}\,d\varsigma - \frac{m}{2(m+n)}\iiint U\rho S\,\nabla\mathbf{F}\,d\varsigma + \text{ the surf. int.}\end{aligned}$$

for by equation (30) $S\,\nabla\mathbf{F} = (m+n)\nabla^2 S\,\nabla\eta$. Now (in order to get rid of any infinite terms due to any discontinuity in **F**) apply equation (9) § 6 to the second volume integral. Thus

$$4\pi n\eta = \iiint u\mathbf{F}\,d\varsigma + \frac{m}{2(m+n)}\iiint S\,\mathbf{F}\nabla\,.\,U\rho\,d\varsigma + \text{ a surf. int.}$$

*[Note added, 1892. For a neater quaternion treatment of this problem see *Phil. Mag.* June, 1892, p. 493.]

The surface integral may be neglected as we may thus verify. Call the volume integral $4\pi n\eta'$. Thus

$$4\pi n\nabla^2\eta' = 4\pi\mathbf{F} + \frac{m}{2(m+n)}\iiint S\mathbf{F}\nabla\,.\,\nabla^2 U\rho\, d\mathfrak{s}$$

$$4\pi m\nabla S\,\nabla\eta' = \frac{m}{n}\iiint S\mathbf{F}\nabla\,.\,\nabla u\, d\mathfrak{s} + \frac{m^2}{2n(m+n)}\iiint S\mathbf{F}\nabla\,.\,\nabla S\,\nabla U\rho\, d\mathfrak{s},$$

so that putting $\nabla U\rho = -2u$ we get

$$n\nabla^2\eta' + m\nabla S\,\nabla\eta' \equiv \mathbf{F},$$

whence we have as a particular solution of equation (30) $\eta = \eta'$ or

$$\eta = \frac{1}{4\pi n}\iiint u\mathbf{F}\, d\mathfrak{s} + \frac{m}{8\pi n(m+n)}\iiint S\mathbf{F}\nabla\,.\,U\rho\, d\mathfrak{s}. \tag{32}$$

This is generally regarded as a solution of the statical problem for an infinite isotropic body. In this case some law of convergence must apply to **F** to make these integrals convergent. Thomson and Tait (*Nat. Phil.* § 730) say that this law is that $\mathbf{F}r$ converges to zero at infinity. This I think can be disproved by an example. Put* $\mathbf{F}r = r^{-a}\lambda$ where λ is a constant vector and a a positive constant less than unity. Equation (32) then gives for the displacement at the origin due to the part of the integral extending throughout a sphere whose centre is the origin and radius R

$$\eta = \frac{m+3n}{3n(m+n)}\frac{R^{1-a}}{1-a}\lambda.$$

Putting $R = \infty$, η also becomes ∞. The real law of convergence does not seem to me to be worth seeking as the practical utility of equation (32) is owing to the fact that it is a particular integral.

The present solution of the problem has only to be compared with the one in Thomson and Tait's *Nat. Phil.* §§ 730–1 to see the immense advantage to be derived from Quaternions.

It is easy to put our result in the form given by them. We have merely to express $S\mathbf{F}\nabla\,.\,U\rho$ in terms of **F** and $r^2 S\mathbf{F}\nabla\,.\,\nabla u$ where r is put for the reciprocal of u. Noting that

$$\nabla u = -u^3\rho, \quad U\rho = u\rho, \quad S\mathbf{F}\nabla\,.\,\rho = -\mathbf{F},$$

*[Note added, 1892. This is not legitimate since it makes $\mathbf{F} = \infty$ for $r = 0$. The reasoning is rectified in the *Phil. Mag.* paper just referred to by putting $\mathbf{F} = 0$ from $r = 0$ to $r = b$ and $\mathbf{F}r = r^{-a}\lambda$ from $r = b$ to $r = \infty$.]

we have at once

$$S\mathbf{F}\nabla \,.\, U\rho = -u\mathbf{F} - u^3\rho S\mathbf{F}\rho$$
$$S\mathbf{F}\nabla \,.\, \nabla u = u^3\mathbf{F} + 3u^5\rho S\mathbf{F}\rho$$

therefore eliminating $\rho S\mathbf{F}\rho$,

$$S\mathbf{F}\nabla \,.\, U\rho = -u\mathbf{F}2/3 - r^2 S\mathbf{F}\nabla \,.\, \nabla u/3,$$

$$\therefore \qquad \eta = \{24\pi n(m+n)\}^{-1}\iiint d\varsigma\,\{2(2m+3n)u\mathbf{F} - mr^2 S\mathbf{F}\nabla \,.\, \nabla u\}, \qquad (33)$$

which is the required form.

23. Calling the particular solution η' as before and putting

$$\eta = \eta' + \eta''$$

the statical problem is reduced to finding η'' to satisfy

$$n\nabla^2\eta'' + m\nabla S\nabla\eta'' = 0$$

and the surface equation either

$$\eta' + \eta'' = \text{ given value,}$$

i.e. $$\eta'' = \text{ given value,}$$

or $$\varpi U d\Sigma = \text{ given surface traction,}$$

i.e. by equation (29) § 21 above,

$$nSU\,d\Sigma\nabla \,.\, \eta'' + n\nabla_1 S\eta_1'' U\,d\Sigma + (m-n)U\,d\Sigma S\nabla\eta'' = \text{ known value.}$$

This general problem for the spherical shell, the only case hitherto solved, I do not propose to work out by Quaternions, as the *methods* adopted are the same as those used by Thomson and Tait in the same problem. But though each step of the Cartesian proof would be represented in the Quaternion, the saving in mental labour which is effected by using the peculiarly happy notation of Quaternions can only be appreciated by him who has worked the whole problem in both notations. The only remark necessary to make is that we may just as easily use vector, surface or solid, harmonics or indeed quaternion harmonics as ordinary scalar harmonics.

Orthogonal coordinates.

24. It is usual to find what equations (22*a*) of § 19 and (3) of § 12 become when expressed in terms of any orthogonal coordinates. This can be done much more easily by Quaternions than Cartesian Geometry. Compare the following investigation with the corresponding one in Ibbetson's *Math. Theory of Elasticity*, Chap. V.

Let x, y, z be any orthogonal coordinates, i.e. let x = const., y = const., z = const., represent three families of surfaces cutting everywhere at right angles. Particular cases are of course the ordinary Cartesian coordinates, the spherical coordinates r, θ, ϕ and the cylindrical coordinates r, ϕ, z. Let D_x, D_y, D_z stand for differentiations *per unit length* perpendicular to the three coordinate surfaces and let λ, μ, ν be the unit vectors in the corresponding directions. Thus

$$\nabla = \lambda D_x + \mu D_y + \nu D_z.$$

Thus, using the same system of suffixes for the D's as was explained in connection with ∇ in § 1,

$$\phi\Delta = D_{x1}\phi_1\lambda + D_{y1}\phi_1\mu + D_{z1}\phi_1\nu, \tag{34}$$

or

$$\phi\Delta = D_x(\phi\lambda) + D_y(\phi\mu) + D_z(\phi\nu) - \phi(D_x\lambda + D_y\mu + D_z\nu). \tag{35}$$

25. Now to put equation (22*a*) § 19 into the present coordinates all that is required is to express $\varpi\Delta$ in terms of those coordinates. Let the coordinates of ϖ be $PQRSTU$. Thus from equation (35) we have

$$\begin{aligned}\varpi\Delta = D_x(P\lambda + U\mu + T\nu) + D_y(U\lambda + Q\mu + S\nu) + D_z(T\lambda + S\mu + R\nu)\\ - \varpi(D_x\lambda + D_y\mu + D_z\nu).\end{aligned}$$

The first thing then is to find $D_x\lambda$, $D_x\mu$, $D_y\lambda$ &c.

Let p_2, p_3 be the principal curvatures normal to x = const., i.e. (by a well-known property of orthogonal surfaces) the curvatures along the lines of intersection of x = const., with z = const., and y = const. p_2, p_3 will be considered positive when* the positive value of dx is on the convex side of the corresponding curvatures. Similarly for $q_3q_1r_1r_2$. Thus for the coordinates r, θ, ϕ;

$$p_2 = p_3 = 1/r, \quad q_3 = \cot\theta/r, \quad q_1 = r_1 = r_2 = 0.$$

*[Note added, 1892. This is contrary to the usual convention.]

Again for r, ϕ, z; $p_2 = 1/r$ and the rest are each zero.

With these definitions we see geometrically that

$$D_x\lambda = -\mu q_1 - \nu r_1, \quad D_x\mu = \lambda q_1, \quad D_x\nu = \lambda r_1. \tag{36}$$

Similarly for $D_y\lambda$, $D_y\mu$, $D_z\lambda$ &c. Thus

$$\begin{aligned}\varpi\Delta &= \lambda(D_xP + D_yU + D_zT) + \mu(\ ) + \nu(\ )\\ &\quad + \lambda(-Qp_2 - Rp_3 + Tr_1 + Uq_1) + \mu(\ ) + \nu(\ )\\ &\quad + \lambda\{(p_2 + p_3)P + (q_3 + q_1)U + (r_1 + r_2)T\} + \mu\{\ \} + \nu\{\ \},\end{aligned}$$

or

$$\begin{aligned}\varpi\Delta &= \lambda\{D_xP + D_yU + D_zT + P(p_2 + p_3) - Qp_2 - Rp_3\\ &\quad + T(2r_1 + r_2) + U(q_3 + 2q_1)\} + \mu\{\ \} + \nu\{\ \}.\end{aligned} \tag{37}$$

26. The other chiefly useful thing in transformation of coordinates in the present subject is the expression for the strain function ψ in terms of the coordinates of displacement. Let u, v, w be these coordinates. Now by equation (3) § 12, remembering (§ 18) that $\psi = \overline{\chi} - 1$ we have

$$-2\psi\omega = S\omega\nabla\,.\,\eta + \nabla_1 S\omega\eta_1,$$

whence
$$-2\psi\lambda = \lambda S\lambda D_x\eta + \mu S\lambda D_y\eta + \nu S\lambda D_z\eta - D_x\eta.$$

But
$$\begin{aligned}D_x\eta &= \lambda D_xu + \mu D_xv + \nu D_xw + uD_x\lambda + vD_x\mu + wD_x\nu\\ &= \lambda(D_xu + vq_1 + wr_1) + \mu(D_xv - uq_1) + \nu(D_xw - ur_1).\end{aligned}$$

Similarly

$$\begin{aligned}D_y\eta &= \lambda(D_yu - vp_2) + \mu(D_yv + wr_2 + up_2) + \nu(D_yw - vr_2)\\ D_z\eta &= \lambda(D_zu - wp_3) + \mu(D_zv - wq_3) + \nu(D_zw + up_3 + vq_3)\\ \therefore\quad 2\psi\lambda &= 2\lambda(D_xu + vq_1 + wr_1) + \mu(D_yu + D_xv - uq_1 - vp_2)\\ &\quad + \nu(D_zu + D_xw - ur_1 - wp_3).\end{aligned} \tag{38}$$

But with Thomson and Tait's notation for pure small strain

$$2\psi\lambda = 2\lambda e + \mu c + \nu b,$$

$$\therefore \quad \left.\begin{aligned} {}^{*}e &= D_x u + vq_1 + wr_1 \\ f &= D_y v + wr_2 + up_2 \\ g &= D_z w + up_3 + vq_3 \\ a &= D_y w + D_z v - wq_3 - vr_2 \\ b &= D_z u + D_x w - ur_1 - wp_3 \\ c &= D_x v + D_y u - vp_2 - uq_1. \end{aligned}\right\} \qquad (39)$$

Thus we have $efgabc$ in terms of the displacement and we have already in equation (24) § 20, which expresses ϖ in terms of ψ, found the values of $PQRSTU$ in terms of e, &c. Finally the expression in equation (37) for $\varpi\Delta$ gives us the equations of equilibrium in terms of P, &c. Thus we have all the materials for considering any problem with the coordinates we have chosen.

All these results can be at once applied to spherical and cylindrical coordinates, but as this has nothing to do with our present purpose—the exemplification of Quaternion methods—we leave the matter here.

Let us as an example of particular coordinates to which this section forms a suitable introduction consider St Venant's Torsion Problem by means of cylindrical coordinates.

Saint-Venant's Torsion Problem.

27. In this problem we consider the equilibrium of a cylinder with any given cross-section, subjected to end-couples, but to no bodily forces and no stress on the curved surface.

We shall take r, ϕ, z as our coordinates, the axis of z being parallel to the generating lines of the cylinder. Let λ, μ, ν be the unit vectors in the directions of $dr, d\phi, dz$ respectively and let

$$\eta = u\lambda + v\mu + w\nu$$

as before.

We shall follow Thomson and Tait's lines of proof—i.e. we shall first find the effect of a simple torsion and then add another displacement and so try to get rid of stress on the curved surface.

---

*[Note added, 1892. In the *Phil. Mag.* June, 1892, p. 488, there is a mistake in the equation just preceding equation 31 and there are two mistakes in equation 31. In the first of these $I(2D_\xi u - {}_\eta\varpi_\zeta v - {}_\zeta\varpi_\eta w)$ should be $2I(D_\xi u - {}_\eta\varpi_\zeta v - {}_\zeta\varpi_\eta w)$. In equation (31) all the 2's should be dropped.]

Holding the section $z = 0$ fixed let us give the cylinder a small torsion of magnitude τ, i.e. let us put

$$\eta = \tau z r \mu, \tag{40}$$

for all points for which τz is small.

The practical manipulation of such expressions as this is almost always facilitated by considering the general value of $Q(\nabla_1, \eta_1)$ where Q is any function linear in each of its constituents. Thus in the present case

$$Q(\nabla_1, \eta_1) = \tau\{zQ(\lambda, \mu) - zQ(\mu, \lambda) + rQ(\nu, \mu)\}.$$

[If Q is symmetrical in its constituents, e.g. in the case of stress below this reduces to the simple form $Q(\nabla_1, \eta_1) = \tau r Q(\nu, \mu)$.] From this we at once see that η satisfies the equation of internal equilibrium

$$n\nabla^2\eta + m\nabla S\nabla\eta = 0,$$

for putting $Q(\alpha, \beta) = \alpha\beta$

$$\nabla\eta = \tau(2z\nu - r\lambda) = \tau\nabla(z^2 - r^2/2)$$

so that both $S\nabla\eta$ and $\nabla^2\eta = 0$.

Again the value for Q at once gives us the stress for

$$\varpi\omega = -nS\omega\nabla \,.\, \eta - n\nabla_1 S\omega\eta_1 - (m - n)\omega S\nabla\eta,$$

or

$$\varpi\omega = -n\tau r(\mu S\omega\nu + \nu S\omega\mu), \tag{41}$$

which is a shearing stress $n\tau r$ on the interfaces perpendicular to μ and ν.

Putting $\omega =$ the unit normal of the curved surface we have for the surface traction

$$\varpi\omega = -n\tau r\nu S\omega\mu.$$

In the figure let the plane of the paper be $z = 0$, O the origin, P a point on the curved surface and OM the perpendicular from O on the tangent at P. Thus $-rS\omega\mu = OP\cos OPM = PM$, PM being reckoned positive or negative according as it is in the positive or negative direction of rotation round Oz. Thus we see that the surface traction is parallel to Oz and $= n\tau PM$.

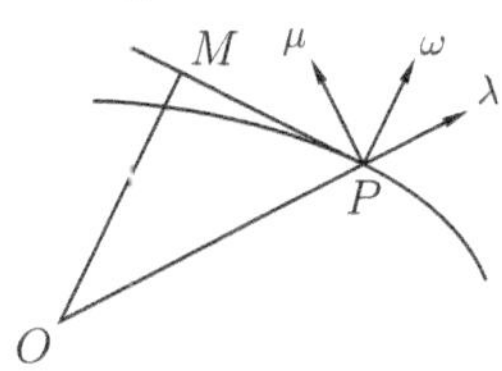

Hence in the case of a circular cylinder a torsion round the axis satisfies all the conditions of our original problem, but this is true in no other case.

The surface traction at any point on the plane ends necessary to produce this strain is $\varpi\nu = n\tau r\mu$ by equation (41) so that its moment round the origin is $n\tau\iint r^2\, dA$, where dA is an element of area and the integral extends over the whole cross-section.

28. Let us now assume a further displacement

$$\eta = w\nu, \tag{42}$$

where w is a function of r, ϕ only, and let us try to determine w so that there is still internal equilibrium and so that the stress on the curved surface due to w shall neutralise the surface traction already considered.

In the present case

$$Q(\nabla_1, \eta_1) = Q(\nabla w, \nu).$$

Thus $S\nabla\eta = 0$ (since ν is perpendicular to ∇w) and therefore the equation of internal equilibrium gives

$$\nabla^2 w = 0.$$

Again
$$\varpi\omega = -nS\,\omega\nabla\,.\,\eta - n\nabla_1 S\,\omega\eta_1,$$
or
$$\varpi\omega = -n(\nu S\,\omega\nabla w + \nabla w S\,\omega\nu), \tag{43}$$

a shear $= nT\nabla w$ on the interfaces perpendicular to ∇w and ν. Thus putting ω for the unit normal to the curved surface, the present surface traction will neutralize the former if

$$S\,\omega\nabla w = -\tau r S\,\omega\mu = -\tau S\,\omega\nu(\lambda r)$$
$$= \tfrac{1}{2}\tau S\,\nu\omega\nabla(r^2),$$
i.e.
$$dw/dn = d(\tau r^2/2)/ds,$$

where d/dn represents differentiation along the normal outwards and d/ds differentiation along the positive direction of the bounding curve.

We leave the problem here to the theory of complex variables and Fourier's Theorem.

Observe however that the surface traction at any point on the plane end is equal to $\varpi\nu = n\nabla w$ by equation (43), and therefore that the total couple is equal to $-n\iint rS\mu\nabla w\, dA = n\iint (dw/d\phi)\, dA$. This leads to the usual expression for torsional rigidity.

Wires.

29. In the following general treatment of Wires some of the processes are merely Thomson and Tait's translated into their shorter Quaternion forms; others are quite different. The two will be easily distinguished by such as are acquainted with Thomson and Tait's *Nat. Phil.*

The one thing to be specially careful about is the notation and its exact meaning. This meaning we give at the outset.

The wires we consider are not necessarily naturally straight but we assume some definite straight condition of the wire as the "geometrically normal" condition.

The variable in terms of which we wish to express everything is s the distance along the wire from some definite point on it.

Any element of the wire, since it is only *slightly* strained, may be assumed to have turned as a rigid body from its geometrically normal position. This rotation is expressed as usual (Tait's *Quaternions*, § 354) by the quaternion q; the axis of q being the axis of rotation, and the angle of q, half the angle turned through.

ω is taken so that the rate of this turning per unit length of the wire is $q\omega q^{-1}$ so that ω is the rate of turning per unit of length when the whole wire is moved as a rigid body so as to bring the element under consideration back to its geometrically normal position. Of course ω is a function of q and its derivative with reference to s. This function we shall investigate later. The resolved part of $q\omega q^{-1}$ parallel to the wire is the vector twist and the resolved part perpendicular to the wire is in the direction of the binormal and equal to the curvature. In fact ω is the vector whose coordinates are the κ, λ, τ of Thomson and Tait's *Nat. Phil.* § 593. When we are given q or ω for every point we know the strain of the wire completely. ω_0 is defined as the naturally normal value of ω, i.e. the value of ω when the wire is unstressed.

As usual we take ρ as the coordinate vector of any point of the wire, ρ like the rest of the functions being considered as a function of s. We shall denote (after Tait, *Quaternions*, Chap. IX.) differentiations with regard to s by dashes.

We now come to the dynamical symbols. **F** and **M** are the force and couple respectively exerted across any normal section of the wire on the part of the wire which is on the negative side of the section by the part on the positive side.

Finally let **X**, **L** be the external force and couple per unit length exerted upon the wire.

30. When the wire is strained in any way let us impose a small additional strain represented by an increment $\delta\omega$ in ω and an increment δe in the elongation at any point. Then the work done on the element ds by the stress-force $= -\delta e S \mathbf{F}\rho' \, ds$ and that done by the stress-couple $= -S q\delta\omega q^{-1}\mathbf{M}\, ds$. If (as we assume, though the assumption is not justified in some useful applications of the general theory of wires) $\mathbf{F}$ and $\mathbf{M}$ to be of the same order of magnitude the former of these expressions can be neglected in comparison with the latter for δe is a quantity small compared with $\delta\omega$. Now the work done on the element by the stress = the increment of the element's potential energy = $\delta w ds$ where w is some function of the strain. Hence

$$\delta w = -S\ \delta\omega\, q^{-1}\mathbf{M}q.$$

Thus w is a function of ω only and

$$\therefore \qquad \delta w = -S\ \delta\omega\, {}_{\omega}\nabla w,$$

whence we see that

$$q^{-1}\mathbf{M}q = {}_{\omega}\nabla w. \tag{46}$$

Notice that $\mathbf{M}$ is Thomson and Tait's ξ, η, ζ and $q^{-1}\mathbf{M}q$ their KLM (*Nat. Phil.* §§ 594, 614).

31. Now since the strain is small, $q^{-1}\mathbf{M}q$ is linear in terms of the strain and therefore in terms of ω. Hence we see that w is quadratic in terms of ω. Let us then put

$$w = w_2(\omega - \omega_0, \omega - \omega_0) + w_1(\omega - \omega_0) \\ + \text{ a function of temperature only,}$$

w_2 and w_1 being linear and homogeneous in each of their constituents. This is the most general quadratic function of ω. Now

$${}_{\omega}\nabla w = \zeta w_2(\zeta, \omega - \omega_0) + \zeta w_2(\omega - \omega_0, \zeta) + \zeta w_1 \zeta.$$

Putting then $\omega = \omega_0$ and ${}_{\omega}\nabla w = 0$ we get $\zeta w_1 \zeta = 0$. Operating on the last by $S\sigma(\ )$ where σ is any vector we see that $w_1 = 0$. Thus putting

$${}_{\omega}\nabla w = \phi(\omega - \omega_0),$$

where ϕ has the value given by the last equation and is therefore self-conjugate we get the two following equations

$$w = -S(\omega - \omega_0)\phi(\omega - \omega_0)/2 + w_0, \tag{47}$$

[as can be seen by a comparison of the last three equations w_0 being the function of the temperature] and

$$q^{-1}\mathbf{M}q = \phi(\omega - \omega_0). \tag{48}$$

When the natural shape of the wire is straight these become

$$w = -S\,\omega\phi\omega/2 + w_0, \tag{49}$$

$$q^{-1}\mathbf{M}q = \phi\omega, \tag{50}$$

and when further the wire is truly uniform ϕ and w_0 are constant along the wire.

32. Assuming the truth of these restrictions let us conceive a rigid body moving about a fixed point which, when placed in a certain position which we shall call the normal position, has, if then rotating with any vector angular velocity ω, a moment of momentum = $\phi\omega$ where ϕ has the meaning just given. If the rigid body be made to take a finite rotation $q(\;)q^{-1}$ and then to move with angular velocity $q\omega q^{-1}$ its moment of momentum will be $q\phi\omega q^{-1}$. Now let a point move along the wire with unit velocity and let the rigid body so move in unison with it that when the moving point reaches the point s the rigid body shall have made the rotation represented by q (§ 29 above). Thus by the definition of ω and q its angular velocity at any instant is $q\omega q^{-1}$ and its moment of momentum therefore $q\phi\omega q^{-1}$ or $\mathbf{M}$.

Now consider the equilibrium of the wire when no external force or couple acts except at its ends. In this case $\mathbf{F}$ is constant throughout and it is easy to see (what indeed is a particular case of equation (53) below) that

$$\mathbf{M}' + V\rho'\mathbf{F} = 0. \tag{51}$$

Interpreting this equation for our rigid body we get as the law which governs its motion

$$d(\text{vect. mom. of mom.})/dt = -V\rho'\mathbf{F}.$$

Thus the rigid body will move as if acted upon by a constant force $\mathbf{F}$ at the end of the unit vector ρ' or—since this vector is fixed in the body—as if acted upon by a constant force acting through a point fixed in the body. From this kinetic analogue of Kirchhoff's the mathematical problem of the shape of such a wire as we are now considering, under the given circumstances, is shewn to be identical with the general problem of the pendulum of which the top is a variety.

33. We will now give the general equations for any wire under any external actions. The comparison of the Quaternion treatment of this with the Cartesian as given in Thomson and Tait's *Nat. Phil.* § 614 seems to me to be all in favour of the former.

The equations of equilibrium of an element ds are with the notation explained in § 29 above

$$d\mathbf{F} + \mathbf{X}ds = 0,$$
$$Vd\rho\mathbf{F} + d\mathbf{M} + \mathbf{L}\,ds = 0,$$

or dividing by ds

$$\mathbf{F}' + \mathbf{X} = 0, \tag{52}$$
$$V\rho'\mathbf{F} + \mathbf{M}' + \mathbf{L} = 0. \tag{53}$$

Operating on the last equation by $V\rho'(\,)$ noting that $\rho'^2 = -1$ and putting $S\rho'\mathbf{F} = -T$ we get

$$\mathbf{F} = \rho'T + V\rho'(\mathbf{M}' + \mathbf{L}), \tag{54}$$

whence by equations (52) and (53) respectively

$$\mathbf{X} + d\{\rho'T + V\rho'(\mathbf{M}' + \mathbf{L})\}/ds = 0, \tag{55}$$
$$S\rho'(\mathbf{M}' + \mathbf{L}) = 0. \tag{56}$$

Now by equation (48) above

$$q^{-1}\mathbf{M}q = \phi(\omega - \omega_0). \tag{48}$$

Also by the definition of q

$$\rho' = q\lambda q^{-1}, \tag{57}$$

where λ is some given constant unit vector. Finally as we are about to prove

$$\omega = 2Vq^{-1}q'. \tag{58}$$

It is usual in the Cartesian treatment to leave the problem in the form of equations equivalent to the above (55) to (58), 13 scalar equations for the 13 unknown scalars of ρ, T, $\mathbf{M}$, q and ω. We can however as we shall directly in the Quaternion treatment quite easily reduce the general problem to one vector and one scalar equation involving the four unknown scalars of T and q in terms of which all the other unknowns are explicitly given.

To prove equation (58)* observe that

$$(q + dq)\sigma(q + dq)^{-1} = q(\sigma + V\omega\sigma\, ds)q^{-1},$$

where σ is any vector. The truth of this is seen by noticing that $(q+dq)(\ )(q+dq)^{-1}$ is the operator that rotates any vector of the element $s + ds$ from its geometrically normal position to its strained position. But we can also get to this final position by first in the geometrically normal wire making the small strain $\omega\, ds$ at the given element and then performing the strain of the wire up to the point s. The first process is represented on the left of the last equation and the second on the right. Thus we get

$$q^{-1}\, dq\, \sigma + \sigma d(q^{-1})\,.\,q = V\omega\sigma\, ds,$$

or $\because$

$$d(q^{-1}) = -q^{-1}\, dq\, q^{-1},$$

$$q^{-1}q'\sigma - \sigma q^{-1}q' = V\omega\sigma,$$

whence

$$\omega = 2V\, q^{-1}q'.$$

Returning to equations (55) to (58) observe that equations (57) and (58) give ρ' and ω as explicit functions of q. Hence by equation (48)

$$\mathbf{M} = q\phi(2V\, q^{-1}q' - \omega_0)q^{-1}, \tag{59}$$

which gives $\mathbf{M}$ also explicitly. Substituting for $\mathbf{M}$ and ρ' in equations (55) and (56) we have

$$\mathbf{X} + \frac{d}{ds}\left\{q\lambda q^{-1}\, T + V\, q\lambda q^{-1}\Big(\mathbf{L} + \frac{d}{ds}\big[q\,\phi(2V\, q^{-1}q' - \omega_0)\, q^{-1}\big]\Big)\right\} = 0, \tag{60}$$

$$S\, q\lambda q^{-1}\Big(\mathbf{L} + \frac{d}{ds}\big[q\,\phi(2V\, q^{-1}q' - \omega_0)\, q^{-1}\big]\Big) = 0, \tag{61}$$

which are sufficient equations to determine T and q, whereupon $\mathbf{M}$ is given by equation (59), ω by equation (58) and ρ' and therefore also ρ by equation (57).

Spiral springs can be treated very simply by means of the above equations, but we have already devoted sufficient space to this subject.

---

*This could be deduced from Tait's *Quaternions*, § 356, equation (2). His ϵ is our $q\omega q^{-1}$ and his dots our dashes.

SECTION IV.

ELECTRICITY AND MAGNETISM.

ELECTROSTATICS

General Problem.

34. Merely observing that all the theorems in integration given in the Preliminary and IVth chapters of Maxwell's treatise on *Electricity and Magnetism*, Part I., are easy particular cases of equations (8) and (9) § 6 above, we pass on to the one application of Quaternions that we propose to make in Electrostatics.

This is to find the most general mechanical results arising from Maxwell's theory of Electrostatics, and to see if they can be explained by stress in the dielectric. This problem as far as I am aware has not hitherto in all its generality been attacked though the most important practical cases have been, as we shall see, considered by Maxwell, Helmholtz, Korteweg, Lorberg and Kirchhoff.

It is necessary first of all to indicate as clearly as possible what I take to be Maxwell's theory of Electricity.

He assumes* all space to be uniformly filled with a certain substance called Electricity. Whatever electrical actions take place depend on the continued or past

*Prof. J. J. Thomson in his paper on *Electrical Theories*, B.A. Reports, 1885, p. 125, does not credit Maxwell with such a definite and circumscribed theory as that described in the text, and he is thereby led to find fault with Maxwell's term "Displacement" and points out that there is an assumption made with reference to the connection between the true current and this polarisation (displacement). He says moreover, "It is rather difficult to see what is meant in Maxwell's Theory by the phrase 'Quantity of Electricity.' " None of these remarks are called for if the view I take of Maxwell's theory be correct, and these grounds alone I consider sufficient for taking that view. The paper of Thomson's here mentioned I shall frequently have to refer to. [Note added, 1892. In the text I have given much too rigid a form to Maxwell's theory. What I have called his theory I ought rather to have called his analogy. Still I think the present foot-note is in the main just. In my opinion it is no more and no less difficult to see what is meant in Maxwell's Theory by "Quantity of Electricity" than by "displacement" since the two are connected by perfectly definite equations. Of course it is wrong to define "displacement" as "displacement of quantity of electricity" and then to define "quantity of electricity" in terms of "displacement" but Maxwell does not seem to me even tacitly to do this. Rather he says—the dielectric is polarised; this polarisation can be

motion of this substance as an incompressible fluid. If electricity is brought from a distance by any means and placed in a given space there must be a displacement of the original electricity outwards from that space and the quantity of foreign electricity is conveniently measured by the surface integral of that displacement.

Dielectrics are substances in which this displacement tends to undo itself, so to speak, i.e. the original electricity tends to go back to its primitive position. In *conductors*, on the other hand, there is no property distinguishing any imported electricity from the original electricity.

The rate of variation of displacement, whether in dielectrics or conductors, of course constitutes an electric current as it is conveniently called.

We have next to consider a vector at each point of space called the electro-motive force, which depends in some way at present undefined on the distribution of the displacement in the dielectrics, the distribution of currents whether in dielectrics or conductors, and on extra-electrical or semi-electrical action, e.g. chemical or mechanical.

If at any point the electro-motive force be multiplied by a scalar the medium at the point remaining (except electrically) unchanged, the current in the case of conductors and the displacement in the case of dielectrics is altered in the same ratio. In other words the current or the displacement, as the case may be, is a linear vector function of the electro-motive force, and the coordinates of the linear vector function* at any point depend solely on the state of the medium (whether fluid, solid, &c., or again strained or not) at that point.

To complete the theory we have to explain how the part of the electro-motive force which is a function of the distribution of displacement and current depends on this distribution. This explanation is obtained by making the assumption that the electro-motive force bears to electricity defined as above exactly the same

---

represented by a vector **D**; electrical quantity can be expressed in terms of **D**; the mathematical connections between electrical quantity and **D** are the same as those between quantity of matter in a space and the displacement out of that space made by other matter to make room for the given matter; we will impress this useful analogy firmly on our minds by calling **D** the displacement. But I have expressed my present views on the meaning of Maxwell's theory much more fully in *Phil. Trans.* 1892, p. 685.]

*This frequently recurring cumbrous mode of description must be tolerated unless a single word can be invented for "a linear vector function of a vector." Might I suggest the term "Hamiltonian?" Thus we should say that the displacement is a Hamiltonian of the electro-motive force, the Hamiltonian at any point being a function of the state of the medium.

energy relation as ordinary force does to matter, i.e.—

(wk. dn. on electricity moved during any displacement)
= (total displacement of elect.)
× (resolved part of E.M.F. in the direction of displacement).

In the IVth part of Maxwell's treatise he gives complete investigations of the mechanical results flowing from this theory so far as it refers to currents, but he has not given the general results in the case of Electrostatics. Nor has he shewn satisfactorily, it seems to me, that the ordinary laws of Electrostatics flow from his theory. It is these investigations we now propose to make.

35. Our notation will be as far as possible the same as Maxwell's. Thus for the displacement at any point we use **D**, and for the E.M.F. **E**. From the connection explained in last section between **D** and **E** we have

$$\mathbf{D} = K\mathbf{E}/4\pi, \tag{1}$$

where K at any point is some linear vector function depending on the state of the medium at the point. If the medium change in any manner not electrical, e.g. by means of ordinary strain K will in general also suffer change.

Let w be the potential energy per unit volume due to the electrical configuration. Thus if a small increment $\delta\mathbf{D}$ be given to **D** at all points, the increment $\iiint \delta w \, d\mathfrak{s}$ in $\iiint w \, d\mathfrak{s}$, the potential energy of the electrical configuration in any space, = work done on the electricity in producing the change,

$$\therefore \qquad \iiint \delta w \, d\mathfrak{s} = -\iiint S\,\mathbf{E}\,\delta\mathbf{D}\, d\mathfrak{s},$$

by the relation stated in § 34 existing between **E** and **D**. Thus limiting the space to the element $d\mathfrak{s}$

$$\delta w = -S\,\mathbf{E}\,\delta\mathbf{D}. \tag{2}$$

Now suppose $\mathbf{D} = n\mathbf{D}'$ so that by equation (1) $\mathbf{E} = n\mathbf{E}'$ where $\mathbf{E}'$, $\mathbf{D}'$ are corresponding E.M.F. and displacement respectively. Thus

$$\delta w = -n\,\delta n\, S\,\mathbf{E}'\mathbf{D}'.$$

Integrating from $n = 0$ to $n = 1$, and finally changing $\mathbf{D}'$, $\mathbf{E}'$ into **D**, **E** we get

$$w = -S\mathbf{DE}/2. \tag{3}$$

From this we get

$$\delta w = -S\,\mathbf{E}\,\delta\mathbf{D}/2 - S\,\mathbf{D}\,\delta\mathbf{E}/2,$$

so that by equation (2) $S\,\mathbf{E}\,\delta\mathbf{D} = S\,\mathbf{D}\,\delta\mathbf{E}$ or by equation (1)

$$S\,\mathbf{E}K\,\delta\mathbf{E} = S\,\delta\mathbf{E}\,K\mathbf{E}.$$

Hence (because $\mathbf{E}$ and $\delta\mathbf{E}$ are quite arbitrary) K is self-conjugate and therefore involves only six instead of nine coordinates*.

In electrostatics the line integral of $\mathbf{E}$ round any closed curve must be zero, for otherwise making a small conductor coincide with the curve we shall be able to maintain a current by § 34, and so (by the same section) constantly do work on it (i.e. as a matter of fact create heat) without altering the statical configuration. Hence $\mathbf{E}$ must have a potential, say v. Thus

$$\mathbf{E} = -\nabla v. \tag{4}$$

Since in an electrostatic field there is no current in a conductor, $\mathbf{E} = 0$ throughout any such conductor and therefore v = const.

36. The charge in any portion of space is defined as the amount of foreign electricity within that space. Thus the charge in any space is the surface integral of the displacement outwards. Thus if there be a charge on the element $d\Sigma$ of a surface in the dielectric this charge = $S\,d\Sigma_a\,\mathbf{D}_a + S\,d\Sigma_b\,\mathbf{D}_b$ where a, b denote the two faces of the element (so that $d\Sigma_a = -d\Sigma_b$) and in accordance with § 1 above $d\Sigma_a$ points away from the region in which the displacement is $\mathbf{D}_a$. Thus σ being the surface density

$$\sigma = [S\,\mathbf{D}U\,d\Sigma]_{a+b}, \tag{5}$$

where the notation $[\ ]_{a+b}$ is used for $[\ ]_a + [\ ]_b$. Similarly if there be finite volume density of foreign electricity, i.e. finite volume density of charge in any space, the charge = $-\iint S\,\mathbf{D}d\Sigma = -\iiint S\,\nabla\mathbf{D}\,d\mathfrak{s}$, so that if D be the volume density

$$D = -S\,\nabla\mathbf{D}. \tag{6}$$

[The reason for having $+S\,d\Sigma_a\,\mathbf{D}_a$ before and $-S\,d\Sigma\,\mathbf{D}$ here is that in the former case we were considering a charge *outside* the region where $\mathbf{D}_a$ is considered—*between* the regions a and b in fact—whereas in the latter case we are considering

*We see from this that $\mathbf{D} = {}_{\mathbf{E}}\nabla w$ or $\mathrm{E} = {}_{\mathbf{D}}\nabla w$ according as w is looked upon as a function of **E** or **D**.

the charge *inside* the region where $\mathbf{D}$ is considered. The same explanation applies to the sign of $-S\mathbf{D}U\,d\Sigma$ for the surface of a conductor given below.]

In conductors, as we saw in § 34, the displacement has virtually no meaning (except when it is changing and so the phenomenon of a current takes place) for the foreign electricity and the original electricity are not to be distinguished. Not so however with the surface of the dielectric in contact with the conductor. We may therefore regard the electricity within the body of the conductor as the original electricity so that the charge is entirely at the surface. Thus the surface density will be $-S\mathbf{D}U\,d\Sigma$ where $d\Sigma$ points away from the conductor and $\mathbf{D}$ is the displacement in the dielectric. This may be regarded as a particular case of equation (5) $\mathbf{D}$ being in accordance with what we have just said considered as zero in the conductor.

37. All the volume integrals with which we now have to deal may be considered either to refer to the whole of space or only to the dielectrics, as the conductors (except at their surfaces) in all cases contribute nothing. The boundary of space will be considered as a surface at infinity and all surfaces where either $\mathbf{D}$ or $\mathbf{E}$ is discontinuous.

Putting $W = \iiint w\,d\mathfrak{s}$ we have already found one expression for W, viz.

$$2W = -\iiint S\mathbf{DE}\,d\mathfrak{s}.$$

We now give another. By equation (4)

$$\begin{aligned}2W &= \iiint S\mathbf{D}\nabla v\,d\mathfrak{s}\\ &= \iint vS\,d\Sigma\mathbf{D} - \iiint vS\nabla\mathbf{D}\,d\mathfrak{s},\end{aligned}$$

by equation (9) § 6 above. Thus by equations (5) and (6) § 36,

$$2W = \iint v\sigma\,ds + \iiint vD\,d\mathfrak{s}, \tag{7}$$

where ds is put, as it frequently will be, for an element of surface, i.e. $T\,d\Sigma$. The value of W which we shall use* is obtained by combining these two, viz.

$$W = \iint v\sigma\,ds + \iiint vD\,d\mathfrak{s} + \tfrac{1}{2}\iiint S\mathbf{DE}\,d\mathfrak{s}. \tag{8}$$

*This is for the general case following the example of Helmholtz in the particular case when K reduces to a single scalar. See *Wiss. Abh.* vol. I. equation (2*d*), p. 805. The *method* adopted in the following investigation is also similar to his.

So far we have merely been shewing that all the above results of Maxwell's flow from what in § 34 has been described as his theory. We now proceed to the actual problem in hand which is proved from these results however they may be obtained. I may remark that *some* such investigation as the above seems to me necessary to make the logic of Maxwell's treatise complete.

38. Suppose now that W is the potential energy of some dynamical system extending throughout space. Let us give to every point of space a small displacement $\delta\eta$ vanishing at infinity and find the consequent increment δW in W. If this can be put in the form

$$\delta W = -\iiint S\,\delta\eta_1\,\phi\nabla_1\,d\varsigma, \tag{9}$$

we shall have the following expression for $\mathbf{F}$ the force per unit volume due to the system

$$\mathbf{F} = \phi\Delta, \tag{10}$$

and the following expression for $\mathbf{F}_S$ the force per unit surface at any surface of discontinuity in ϕ

$$\mathbf{F}_s = -[\phi U\,d\Sigma]_{a+b}, \tag{11}$$

the notation being the same as in equation (5) § 36.

Moreover if ϕ be self-conjugate the forces both throughout the volume and at surfaces of discontinuity are producible by the *stress* ϕ as can be seen by § 13 above. [Compare all these statements with § 14 above.]

For proof, we have by equation (9) § 6

$$\delta W = -\iiint S\,\delta\eta_1\,\phi\nabla_1\,d\varsigma = -\iint S\,\delta\eta\,\phi\,d\Sigma + \iiint S\,\delta\eta\,\phi_1\nabla_1\,d\varsigma,$$

where of course the element $d\Sigma$ is taken *twice*, i.e. once for each face. But

$$\begin{aligned}\delta W &= -(\text{work done by the system } \mathbf{F},\, \mathbf{F}_s \text{ of forces})\\ &= \iint S\,\delta\eta\,\mathbf{F}_s ds + \iiint S\,\delta\eta\,\mathbf{F}\,d\varsigma,\end{aligned}$$

where the element ds is taken only *once*. Equating the coefficients of the arbitrary vector $\delta\eta$ for each point of space we get the required equations (10) and (11).

39. We must then put δW where W is given by equation (8) in the form given in equation (9).

We must first define δ when applied to a function of the position of a point. Suppose by means of the small displacement $\delta\eta$ any point P moves to P'. Then Q

being the value at P, before the displacement $\delta\eta$, of a function of the position of a point, $Q + \delta Q$ is defined as the value of the function at P' after the displacement. Thus even in the neighbourhood of a surface of discontinuity δQ is a small quantity of the same order as $\delta\eta$.

Now the charge within any space, that is the quantity of foreign electricity within that space will not be altered by the strain.

$$\therefore \qquad \delta(D\,d\mathfrak{s}) = 0, \qquad \delta(\sigma\,ds) = 0. \tag{12}$$

To find $\delta\nabla$ we have

$$S\,(d\rho + \delta d\rho)(\nabla + \delta\nabla). = S\;d\rho\,\nabla.$$

or, since $\delta d\rho = -S\;d\rho\,\nabla\,.\,\delta\eta$,

$$S\;d\rho\,\delta\nabla = S\;d\rho\,\nabla_1 S\;\delta\eta_1\,\nabla,$$

whence

$$\delta\nabla = \nabla_1 S\;\delta\eta_1\,\nabla. \tag{13}$$

The part of δW depending on the first two terms of equation (8) is by equation (12)

$$\begin{aligned}&\iiint\delta v\,D\,d\mathfrak{s} + \iint\delta v\,\sigma\,ds\\ &\quad= -\iiint\delta v\,S\,\nabla\mathbf{D}\,d\mathfrak{s} + \iint\delta v\,S\;d\Sigma\,\mathbf{D} &&\text{[by equations (5) and (6) § 36]}\\ &\quad= \iiint S\,\mathbf{D}\nabla\,\delta v\,d\mathfrak{s} &&\text{[by equation (9) § 6].}\end{aligned}$$

Noticing that $\delta\,d\mathfrak{s} = -d\mathfrak{s}\,S\,\nabla\,\delta\eta$ and that $4\pi\delta\mathbf{D} = K\delta\mathbf{E} + \delta K\mathbf{E}$ we see that the last term in equation (8) contributes

$$\iiint S\,\mathbf{D}\,\delta\mathbf{E}\,d\mathfrak{s} - \tfrac{1}{2}\iiint S\,\mathbf{DE}S\,\nabla\,\delta\eta\,d\mathfrak{s} + (8\pi)^{-1}\iiint S\,\mathbf{E}\,\delta K\,\mathbf{E}\,d\mathfrak{s}.$$

Combining the last result with the first term of this we get

$$\begin{aligned}\iiint S\,\mathbf{D}(\nabla\,\delta v - \delta(\nabla v))\,d\mathfrak{s} &= -\iiint S\,\mathbf{D}\,\delta\nabla\,.\,v\,d\mathfrak{s}\\ &= -\iiint S\,\mathbf{D}\nabla_1 S\;\delta\eta_1\,\nabla v\,d\mathfrak{s}\\ &= \iiint S\,\mathbf{D}\nabla_1 S\,\mathbf{E}\,\delta\eta_1\,d\mathfrak{s}.\end{aligned}$$

Thus we have

$$\delta W = -\iiint S\;\delta\eta_1(\tfrac{1}{2}\nabla_1 S\,\mathbf{DE} - \mathbf{E}S\,\nabla_1\mathbf{D})\,d\mathfrak{s} + (8\pi)^{-1}\iiint S\,\mathbf{E}\,\delta K\,\mathbf{E}\,d\mathfrak{s}. \tag{14}$$

40. Now the increment δK in K is caused by two things viz. the mere rotation of the body and the change of *shape* of the body. Let us call these parts δK_r and δK_s respectively.

First consider δK_r. Suppose the rotation is ϵ so that any vector which was ω becomes thereby $\omega + V\epsilon\omega$. Thus the result of operating on $\omega + V\epsilon\omega$ by $K + \delta K_r$ is the same as first operating on ω by K and then rotating. In symbols

$$(K + \delta K_r)(\omega + V\epsilon\omega) = K\omega + V\epsilon K\omega,$$

whence
$$\delta K_r\,\omega = V\epsilon K\omega - K(V\epsilon\omega).$$

Thus
$$S\mathbf{E}\delta K_r\mathbf{E} = S\mathbf{E}\epsilon K\mathbf{E} - S\mathbf{E}K(V\epsilon\mathbf{E})$$
$$= 2S\,\epsilon K\mathbf{E}\mathbf{E} = 8\pi S\,\epsilon\mathbf{D}\mathbf{E},$$

whence giving ϵ its value $V\nabla\delta\eta/2$,

$$(8\pi)^{-1}S\mathbf{E}\,\delta K_r\mathbf{E} = S\nabla\,\delta\eta\,V\mathbf{D}\mathbf{E}/2.$$

Substituting in equation (14)

$$\delta W = -\iiint S\,\delta\eta_1\{V\nabla_1\mathbf{D}\mathbf{E}/2 - \mathbf{E}S\nabla_1\mathbf{D}\}\,d\mathfrak{s} + (8\pi)^{-1}\iiint S\mathbf{E}\,\delta K_s\,\mathbf{E}\,d\mathfrak{s},$$

or

$$\delta W = \tfrac{1}{2}\iiint S\,\delta\eta_1\,V\mathbf{D}\,\nabla_1\mathbf{E}\,d\mathfrak{s} + (8\pi)^{-1}\iiint S\mathbf{E}\,\delta K_s\,\mathbf{E}\,d\mathfrak{s}. \qquad (15)$$

It only remains to consider δK_s. K is a function of the pure strain of the medium and δK_s is the increment in K due to the increment in pure strain owing to $\delta\eta$. Calling this increment of pure strain $\delta\psi$ so that by equation (3) § 12 above

$$\delta\psi\omega = -\delta\eta_1\,S\omega\nabla_1/2 - \nabla_1 S\,\omega\,\delta\eta_1/2,$$

we have

$$\delta K_s = -S\,\delta\psi\,\zeta_\psi\mathfrak{a}_2\zeta\,.\,K_2, \qquad (16)$$

by equation (7) § 5 above. This gives by equation (5) § 3

$$\delta K_s = -S\,\delta\eta_{1\psi}\mathfrak{a}_2\nabla_1\,.\,K_2, \qquad (17)$$

$$\therefore \qquad S\mathbf{E}\,\delta K_s\mathbf{E} = -S\,\delta\eta_{1\psi}\mathfrak{a}_2\nabla_1\,S\mathbf{E}K_2\mathbf{E}. \qquad (18)$$

Now by equations (1) and (3) § 35,

$$8\pi w = -S\mathbf{E}K\mathbf{E},$$

so that w is a function of the independent variables $\mathbf{E}$, ψ (because K is a function of ψ). Therefore

$$(8\pi)^{-1} S\,\mathbf{E}\,\delta K_s\,\mathbf{E} = S\,\delta\eta_{1\psi}\text{Ɑ}w\,\nabla_1. \tag{19}$$

This equation might have been deduced at once thus

$$(8\pi)^{-1} S\,\mathbf{E}\,\delta K_s\,\mathbf{E} = S\,\delta\psi\,\zeta_\psi\text{Ɑ}w\,\zeta = S\,\delta\eta_{1\psi}\text{Ɑ}w\,\nabla_1,$$

but equation (17) is itself of importance so the above proof is preferable.

Thus finally from equation (15) we get

$$\delta W = \iiint S\,\delta\eta_1(V\mathbf{D}\nabla_1\mathbf{E}/2 + {}_\psi\text{Ɑ}w\,\nabla_1)\,d\mathfrak{s}. \tag{20}$$

We therefore have for ϕ in equations (10) and (11) § 38,

$$\phi\omega = -V\mathbf{D}\omega\mathbf{E}/2 - {}_\psi\text{Ɑ}w\,\omega. \tag{21}$$

This is a self-conjugate function so that as we saw in § 38 it is a stress which serves to explain forces both throughout the volume of the dielectric and over any surfaces of discontinuity in $\mathbf{D}$ or $\mathbf{E}$*.

The force in particular cases.

41. Let us first consider that part $-V\mathbf{D}\Delta\mathbf{E}/2$ of the force (equations (10) § 38 and (21) § 40) which does not depend on the variation of K with the shape of the body.

Suppose our dielectric is homogeneous and electrically isotropic so that K is a simple constant scalar. In this case

$$4\pi\mathbf{D} = -K\nabla v, \tag{22}$$

by equations (1) and (4) § 35. Therefore by equations (5) and (6) § 36,

$$4\pi\mathbf{D} = K\nabla^2 v, \tag{23}$$

$$4\pi\sigma = -K[S\,U\,d\Sigma\,\nabla v]_{a+b}. \tag{24}$$

From these we at once get by the theory of potential that

$$Kv = \iiint u\mathbf{D}\,d\mathfrak{s} + \iint u\sigma\,ds. \tag{25}$$

*As far as I am aware nobody has hitherto attempted to find the electrical forces much less the stress except in the case when **D** is parallel to **E** i.e. the dielectric is electrically isotropic when unstrained. The particular results contained in § 45 below have been obtained by Korteweg, Lorberg and Kirchhoff as is stated in Prof. J. J. Thomson's paper (p. 155) referred to in § 34.

From this we know by the theory of potential that at the surface where the charge σ resides ∇v is discontinuous only with regard to its normal component and at all other points is continuous. Thus

$$\nabla v_a = \nabla v_b + xU\,d\Sigma_a$$

and by equation (24) $x = 4\pi\sigma/K$ so that

$$(4\pi)^{-1}K(\nabla v_a - \nabla v_b) = \sigma U\,d\Sigma_a,$$

whence

$$(4\pi)^{-1}KU\,d\Sigma_a(\nabla v_a - \nabla v_b) = -\sigma. \tag{26}$$

Now the force **F** per unit volume is

$$-V\mathbf{D}\Delta\mathbf{E}/2 = -KV\nabla v\Delta\nabla v/8\pi$$
$$= -K\nabla^2 v\nabla v/4\pi,$$

or

$$\mathbf{F} = -\mathbf{D}\nabla v, \tag{27}$$

and the force per unit surface $\mathbf{F}_s$ is by equation (11) § 38,

$$\begin{aligned}\tfrac{1}{2}[V\mathbf{D}U\,d\Sigma\mathbf{E}]_{a+b} &= (8\pi)^{-1}K[V\nabla vU\,d\Sigma\,\nabla v]_{a+b}\\ &= (8\pi)^{-1}KV(\nabla v_aU\,d\Sigma_a\,\nabla v_a - \nabla v_bU\,d\Sigma_a\,\nabla v_b)\\ &= (8\pi)^{-1}KV(\nabla v_a + \nabla v_b)U\,d\Sigma_a\,(\nabla v_a - \nabla v_b),\end{aligned}$$

whence by equation (26)

$$\mathbf{F}_s = -\tfrac{1}{2}\sigma[\nabla v]_{a+b}. \tag{28}$$

Thus we see that Maxwell's theory as given in § 34 above reduces to the ordinary theory when K is a single scalar. In fact two particles containing charges $e\ e'$ apparently repel one another with a force ee'/Kr^2 where r is the distance between them, for by equations (25), (27) and (28) the force in any charged body is that due to a field of potential v given by

$$Kv = \Sigma ue. \tag{29}$$

42. If the medium when strained remain electrically isotropic ${}_{\psi}\daleth K$ as well as K must be a simple scalar. Thus with Thomson and Tait's notation for strain, which makes the coordinates of ψ, e, f, g, $a/2$, $b/2$, $c/2$ we have

$${}_{\psi}\daleth K = dK/de = dK/df = dK/dg$$
$$dK/da = dK/db = dK/dc = 0.$$

Therefore K is a function of $e + f + g$ only, i.e. of the density (m) of the medium. Thus because

$$de + df + dg = -dm/m = -d\log m$$

we get ${}_{\psi}\daleth K = -dK/d\log m = -k$ suppose. Hence

$${}_{\psi}\daleth w = (8\pi)^{-1} S\,\mathbf{E}(dK/d\log m)\mathbf{E} = k\mathbf{E}^2/8\pi.$$

Thus the force $-{}_{\psi}\daleth w\,\Delta$ [equations (21) § 40 and (10) § 38] resulting from the change of K with pure strain is in the case we are now considering

$$-\,k\nabla\mathbf{E}^2/8\pi, \tag{30}$$

and is* therefore, according as k is positive or negative, in the direction of or that opposite to that of the most rapid increase of the square of the electro-motive force. Thus even in the case of a fluid dielectric which has no internal charge but which forms part of a non-uniform field of (electro-motive) force the surfaces of equal pressure and therefore the free surface will if originally plane no longer remain so.

Nature of the Stress.

43. We have seen that the stress which serves to explain the electrostatic forces is that given by equation (21) § 40, viz.

$$\phi\omega = -V\mathbf{D}\omega\mathbf{E}/2 - {}_{\psi}\daleth w\,\omega. \tag{21}$$

Let us first consider the part $-V\mathbf{D}\omega\mathbf{E}/2$ which does not depend on the variation of K. Putting ω first $= U\mathbf{D}$ and then $= U\mathbf{E}$ we get

$$\phi U\mathbf{D} = T\mathbf{D}T\mathbf{E}\,.\,U\mathbf{E}/2,$$
$$\phi U\mathbf{E} = T\mathbf{D}T\mathbf{E}\,.\,U\mathbf{D}/2.$$

Therefore putting ω first = any multiple of $U\mathbf{D} + U\mathbf{E}$ and then = any multiple of $U\mathbf{D} - U\mathbf{E}$ we get

$$\phi\omega = \;\; T\mathbf{D}T\mathbf{E}\omega/2,$$
$$\phi\omega = -T\mathbf{D}T\mathbf{E}\omega/2.$$

*This is the same result as Helmholtz's on the same assumption *Wiss. Abh.* I. p. 798.

Lastly, since

$$-V\mathbf{D}\omega\mathbf{E} = \omega S\,\mathbf{DE} - \mathbf{D}S\,\omega\mathbf{E} - \mathbf{E}S\,\omega\mathbf{D},$$

we see that if we put ω = any multiple of $V\mathbf{DE}$

$$\phi\omega = \omega S\,\mathbf{DE}/2 = -\omega w.$$

Thus we see that the stress now considered is a tension along one of the bisectors of **D** and **E** (the bisector of the positive directions or the negative directions of both) $= T\mathbf{D}T\mathbf{E}/2$, an equal pressure along the other bisector and a pressure $= w$ perpendicular to both these directions. When **D** is parallel to **E** this at once reduces to Maxwell's case, viz. a tension in the direction of **E** and a pressure in all directions at right angles each $= w$.

44. We have now to consider the other part of the stress, viz.

$$\phi\omega = -{}_{\psi}\mathsf{Q}w\,\omega,$$

or

$$\phi\omega = {}_{\psi}\mathsf{Q}_1\omega\,S\,\mathbf{E}K_1\mathbf{E}/8\pi. \tag{31}$$

If we assume that K is a function of the density (m) of the medium only we shall have

$$dK/de = dK/df = dK/dg = -dK/d\log m = -k,$$

say, and

$$dK/da = dK/db = dK/dc = 0,$$

as in § 42. Here however k is not in general a mere scalar but a self-conjugate linear vector function. We have then in this case

$$\phi\omega = -\omega S\,\mathbf{E}k\mathbf{E}/8\pi,$$

which is a hydrostatic pressure or an equal tension in all directions according as $S\,\mathbf{E}k\mathbf{E}$ is positive or negative. In this case the 36 coordinates of ${}_{\psi}\mathsf{Q}_1\omega\,.\,K_1$ reduce to the 6 of k for each point of space.

A more general assumption is that δK_s (§ 40) depends only on the elongations in the directions of the principal axes of K. Taking i, j, k as unit vectors in these directions we again have

$$dK/da = dK/db = dK/dc = 0,$$

and thus

$$\phi i = iS\,\mathbf{E}(dK/de)\mathbf{E}/8\pi,$$

and similarly for j and k, so that the principal axes of the stress now considered are the principal axes of K.

45*. The most natural simple assumption for solid dielectrics seems to me to be that the medium is electrically isotropic before strain, and also isotropic with regard to the strain in the sense that if the strain be, so to speak, merely rotated, δK_s will suffer exactly the same rotation. We may treat this problem exactly as we did (§ 20) that of stress in terms of strain for an isotropic solid. Thus splitting up $\delta\psi$ into its principal elongations, i.e. putting

$$\delta\psi\,\omega = -\Sigma\,\delta ei\,S\,i\omega,$$

we shall get, as in § 20,

$$\begin{aligned}\delta K_s\,\omega &= -(\alpha-\beta)\Sigma\,\delta ei\,S\,i\omega + \beta\omega\Sigma\,\delta e\\ &= (\alpha-\beta)\,\delta\psi\,\omega - \beta\omega S\,\zeta\,\delta\psi\,\zeta.\end{aligned}$$

But $\delta K_s\omega = -S\;\delta\psi\,\zeta_\psi \Delta_1\zeta \,.\, K_1\omega$ by equation (16) § 40, so that from equation (31)

$$-8\pi S\;\delta\psi\,\zeta\phi\zeta = (\alpha-\beta)S\,\mathbf{E}\,\delta\psi\,\mathbf{E} - \beta\mathbf{E}^2 S\,\zeta\,\delta\psi\,\zeta,$$

whence we see by § 4 above that

$$\phi\omega = \{(\alpha-\beta)\mathbf{E}S\,\mathbf{E}\omega + \beta\mathbf{E}^2\omega\}/8\pi, \tag{32}$$

which consists of a pressure in the directions of the lines of force $= -\alpha\mathbf{E}^2/8\pi$ and another pressure in all directions at right angles $= -\beta\mathbf{E}^2/8\pi$.

MAGNETISM

Magnetic potential, force, induction.

46. We now go on to the ordinary theory of magnetism; and here we shall merely follow Maxwell in his General Theory, so as to give an opportunity of comparing Quaternion proofs with Cartesian, as we have already done in Elasticity.

We shall not consider in detail the effect of one small magnet upon another, as this has already been done by Tait. In connection with this I am content to remark that I think the treatment of this problem can be made somewhat simpler than Tait's by means of *potential*.

Suppose we have a pole $-m$ at O and a pole $+m$ at O' where OO' is small. Calling the vector from O to O' $\overline{OO'}$, let us call the vector $m\overline{OO'}$ μ, so that μ is

*For references to former proofs of this see foot-note to § 40 above.

the vector magnetic moment of the magnet. The potential of $-m$ at any point P is $-mu$, where u as usual $= PO^{-1}$. Similarly the potential of $+m$ is mu', where $u' = PO'^{-1}$. Therefore the potential of the magnet

$$= m(u' - u) = mS\overline{OO'}\nabla u,$$

where of course P is the variable point implied by ∇. Thus the potential of a small magnet μ at any point $= S\mu\nabla u$.

Hence the potential of *any* magnet whose magnetic moment per unit volume at any point is **I** is

$$\Omega = S\nabla\iiint u\mathbf{I}\,d\mathfrak{s} = -\iiint S\,\mathbf{I}\nabla u\,d\mathfrak{s}, \tag{33}$$

according to the convention of § 9 above. By equation (9) § 6 this may be put

$$\Omega = -\iint uS\,\mathbf{I}\,d\Sigma + \iiint uS\,\nabla\mathbf{I}\,d\mathfrak{s}, \tag{34}$$

which shews that we may consider it due to a volume density $S\nabla\mathbf{I}$ and a surface density $-S\,\mathbf{I}U\,d\Sigma$* of magnetic matter, the surface density occurring wherever there is discontinuity in **I**.

By again considering the poles m and $-m$ of the small magnet μ we see that its potential energy when placed in a field of magnetic potential Ω is $-S\mu\nabla\Omega$, whence just as we obtained equation (33) we now see that the potential energy (W) of any magnet in such a field is

$$W = -\iiint S\,\mathbf{I}\nabla\Omega\,d\mathfrak{s}, \tag{35}$$

or

$$W = -\iint \Omega S\,\mathbf{I}\,d\Sigma + \iiint \Omega S\,\nabla\mathbf{I}\,d\mathfrak{s}, \tag{36}$$

by equation (9) § 6 above, so that the potential energy is just the same as it would be for the imaginary distribution of magnetic matter.

47. The force (**H**) on a unit magnetic pole at any point external to the magnet is given by

$$\begin{aligned}\mathbf{H} = -\nabla\Omega = -\nabla S\nabla\iiint u\mathbf{I}\,d\mathfrak{s} &= -\nabla^2\iiint u\mathbf{I}\,d\mathfrak{s} + \nabla V\nabla\iiint u\mathbf{I}\,d\mathfrak{s}\\ = -4\pi\mathbf{I} + \nabla\iiint V\mathbf{I}\nabla u\,d\mathfrak{s} &= -4\pi\mathbf{I} + \nabla\mathbf{A},\end{aligned}$$

where

$$\mathbf{A} = \iiint V\mathbf{I}\nabla u\,d\mathfrak{s}. \tag{37}$$

*[Note added, 1892. More generally and better $-[S\,\mathbf{I}U\,d\Sigma]_{a+b}$.]

Thus we see that for all external points $\mathbf{H} = \nabla\mathbf{A}$, so that $\mathbf{A}$ is called the vector magnetic potential. [It is to be observed that since $\nabla\mathbf{A} = \mathbf{H} + 4\pi\mathbf{I}$ = a vector, $S\nabla\mathbf{A} = 0$.] $\nabla\mathbf{A}$ is called the magnetic induction and for it we use the single symbol $\mathbf{B}$ so that

$$\mathbf{B} = \nabla\mathbf{A}. \tag{38}$$

Thus

$$S\nabla\mathbf{B} = S\nabla^2\mathbf{A} = 0, \tag{39}$$

and also by the equation for $\mathbf{H}$ just given

$$\mathbf{B} = \mathbf{H} + 4\pi\mathbf{I}. \tag{40}$$

This is not the way in which Maxwell defines the magnetic force and induction, but he shews quite simply (*Elect. and Mag.* §§ 398–9) that his definition and the present one are identical. This can be shewn as easily without analysis at all.

48. Where $\mathbf{I}$ is discontinuous both $\mathbf{H}$ and $\mathbf{B}$ are also discontinuous. From the surface density view we gave in equation (34) we see that, just as we have the expression, given in § 41 for $\nabla v_a - \nabla v_b$, so now

$$\mathbf{H}_b - \mathbf{H}_a = -4\pi U\, d\Sigma_a[S\mathbf{I}U\, d\Sigma]_{a+b}, \tag{41}$$

so that the discontinuity in $\mathbf{H}$ is entirely normal to the surface of discontinuity. Further from this equation we have

$$S(\mathbf{H}_b - \mathbf{H}_a)\, d\Sigma_a = 4\pi[S\mathbf{I}\, d\Sigma]_{a+b},$$

i.e.

$$[S(\mathbf{H} + 4\pi\mathbf{I})\, d\Sigma]_{a+b} = 0,$$

or

$$[S\mathbf{B}\, d\Sigma]_{a+b} = 0 \tag{42}$$

so that the discontinuity in $\mathbf{B}$ is entirely tangential.

From this equation we see that for any closed surface whatever whether it include surfaces of discontinuity in $\mathbf{I}$ or not

$$\iint S\, d\Sigma\, \mathbf{B} = 0.$$

For adding these surfaces to the boundary of the inclosed space, in accordance with § 7 above, we see by equation (42) that they contribute zero to the surface integral; but the total surface integral is by equation (9) § 6 $\iiint S\nabla\mathbf{B}\, d\mathfrak{s} = 0$ by equation (39).

Magnetic Solenoids and Shells.

49. A magnet is said to be solenoidal if the imaginary magnetic matter of equation (34) is entirely on the surface. Thus for a solenoidal distribution

$$S\nabla\mathbf{I} = 0. \tag{43}$$

In this case the potential is by equation (34)

$$\Omega = -\iint uS\mathbf{I}\,d\Sigma. \tag{44}$$

50. A simple magnetic shell is defined as a sheet magnetised everywhere normally to itself and such that, at any point, the magnetic moment per unit surface is a constant called the strength of the shell.

Calling the strength ϕ we have for the potential energy at any point by equation (33)

$$\Omega = -\phi\iint S\,d\Sigma\,\nabla u. \tag{45}$$

Now $-S\,d\Sigma\,\nabla u$ is the solid angle subtended by the element $d\Sigma$ at the point considered, so that

$$\Omega = \phi \times (\text{solid angle subtended by shell at point}). \tag{46}$$

Thus if P be a point on the positive side of the shell and P' a point infinitely near P but on the negative side

$$\text{Potential at } P - \text{Potential at } P' = 4\pi\phi,$$

or what comes to the same thing

$$-\int_P^{P'} S\mathbf{H}\,d\rho = 4\pi\phi.$$

This integral may be taken along any path, e.g. along a path which nowhere cuts the shell. The same integral is true if **H** be the magnetic force due to a whole field of which the shell is only one of several causes, for the part contributed by the rest of the field is zero on account of the infinite proximity of P and P'. For future use in electro-magnetism observe that this statement cannot be made if for **H** in the integral be substituted **B**.

51. The condition that any magnet can be divided up into such shells is at once seen to be that **I** can be put in the form

$$\mathbf{I} = \nabla\phi, \tag{47}$$

where ϕ is some scalar.

In this case the potential is by equation (33)

$$\Omega = -\iiint S\,\mathbf{I}\nabla u\,d\mathfrak{s} = -\iiint S\,\nabla\phi\nabla u\,d\mathfrak{s},$$

or by equation (9) § 6

$$\Omega = -\iint \phi S\ d\Sigma\,\nabla u + \iiint \phi\nabla^2 u\,d\mathfrak{s},$$

i.e.

$$\Omega = -\iint \phi S\ d\Sigma\nabla u + 4\pi\phi. \tag{48}$$

Remarking that the solid angle again occurs here it is needless to interpret the equation further. By equation (37) we have for the vector potential

$$\mathbf{A} = \iiint V\mathbf{I}\nabla u\,d\mathfrak{s} = \iiint V\nabla\phi\nabla u\,d\mathfrak{s},$$

or by equation (9) § 6

$$A = \iint \phi V\,d\Sigma\,\nabla u. \tag{49}$$

52. The potential energy of a magnetic shell of strength ϕ placed in a field of potential Ω is of importance. We see by equation (35) that it is

$$W = -\phi\iint S\ d\Sigma\,\nabla\Omega.$$

If then the magnets which cause Ω do not cut the shell anywhere, so that $-\nabla\Omega = \nabla\mathbf{A}$, we shall have

$$W = \phi\iint S\ d\Sigma\,\nabla\mathbf{A} = \phi\iint S\ d\Sigma\mathbf{B}, \tag{50}$$

or

$$W = \phi\int S\ d\rho\,\mathbf{A}, \tag{51}$$

by equation (8) § 6.

Suppose now that **A** is caused by another shell of strength ϕ'. Then by equation (49)

$$\mathbf{A} = \phi'\iint V\,d\Sigma'\,\nabla u = \phi'\int u\,d\rho',$$

by equation (8) § 6. Thus finally the potential energy M of these two shells is given by

$$M = \phi\phi'\iint uS\ d\rho\,d\rho'. \tag{52}$$

53. The general theory of induced magnetism when once the proposition (given in equation (42) § 48) that $[S\, d\Sigma\mathbf{B}]_{a+b}$ is zero is established, is much the same whether treated by Quaternion or Cartesian notation. We shall therefore not enter into this part of the subject.

Electro-magnetism

General theory.

54. We now propose to prove the geometrical theorems connected with Maxwell's general theory of Electro-magnetism by means of Quaternions.

We assume the dynamical results of Chaps. V., VI. and VII., and the first six paragraphs of Chap. VIII. of the fourth part of his treatise.

These assumptions amount to the following. Connected with any closed curve in an electro-magnetic field there is a function

$$p = -\int S\mathbf{A}\, d\rho, \tag{53}$$

where $\mathbf{A}$ is some vector function at every point of the field. The function p has the following properties. If any circuit be made to coincide with the curve the generalised force acting upon the *electricity* in the circuit is

$$E = -\dot{p}. \tag{54}$$

Again, if there be a current of electricity γ flowing round this circuit, the generalised force X, corresponding to any coordinate x of the position of the circuit due to the field acting upon the *conductor*, is

$$X = \gamma dp/dx. \tag{55}$$

55. The first thing to be noticed is that p can be transformed into a surface integral by equation (8) § 6 above.

Thus $$p = -\iint S\mathbf{B}\, d\Sigma, \tag{56}$$

where $$\mathbf{B} = V\nabla\mathbf{A}, \tag{57}$$

so that $$S\nabla\mathbf{B} = 0.$$

Next we see by the fundamental connection (§ 34 above) between the E.M.F. $\mathbf{E}$ and electricity, that E must equal the line integral of $\mathbf{E}$ round the circuit, or

$$E = -\int S\mathbf{E}\, d\rho. \tag{58}$$

We are now in a position to find $\mathbf{E}$ in terms of $\mathbf{A}$ and $\mathbf{B}$, i.e. of $\mathbf{A}$.

56. The rate of variation of p is due to two causes, viz. the variation of the field ($\dot{\mathbf{A}}$) and the motion of the circuit ($\dot{\rho}$). In the time δt then there will be an increment $\delta\mathbf{A}$ in $\mathbf{A}$ and an increment $\delta d\Sigma$ in $d\Sigma$ to be considered. Thus

$$\dot{p}\delta t = -\int S\,\delta\mathbf{A}\,d\rho - \iint S\,\mathbf{B}\,\delta d\Sigma.$$

[This amounts to assuming that $\int S\,\delta\mathbf{A}\,d\rho = \iint S\,\delta\mathbf{B}\,d\Sigma$, which of course is true by equation (8) § 6.] Now when the circuit changes slightly we may suppose the surface over which the new integral extends to coincide with the original surface and a small strip at the boundary traced out by the motion ($\dot{\rho}\delta t$) of the boundary. Thus $\delta d\Sigma$ is zero everywhere except at the boundary and there it

$$= V(\dot{\rho}\delta t)\,d\rho,$$

so that
$$\dot{p}\,\delta t = -\int S\;\delta\mathbf{A}\,d\rho + \delta t\int S\,\dot{\rho}\mathbf{B}\,d\rho,$$

whence dividing by δt

$$\dot{p} = \int S\,(-\dot{\mathbf{A}} + V\dot{\rho}\mathbf{B})\,d\rho, \tag{59}$$

but by equations (54) and (58) $\dot{p} = \int S\,\mathbf{E}\,d\rho$. Thus

$$\mathbf{E} = -\dot{\mathbf{A}} + V\dot{\rho}\mathbf{B} - \nabla\psi, \tag{60}$$

where ψ is a scalar and $-\nabla\psi$ is put as the most general vector whose line integral round *any* closed curve is zero.

57. We now come to the mechanical forces exerted on an element through which a current $\mathbf{C}$ per unit volume flows.

We see by equation (55) that the work done by the mechanical forces on any circuit through which a current of magnitude γ flows in any small displacement of the circuit equals $\gamma\times$ the increment in p caused by the displacement. Give then to each element $d\rho$ of the circuit an arbitrary small displacement $\delta\rho$ and let $\mathbf{F}'$ be the mechanical force exerted by the field upon the element. Thus as in last section

$$-\int S\mathbf{F}'\,\delta\rho = \gamma\,\delta p = -\gamma\iint S\,\mathbf{B}\,\delta d\Sigma = -\gamma\int S\;\delta\rho\,d\rho\,\mathbf{B}.$$

Thus the force $\mathbf{F}'$ on the element $d\rho$ is $\gamma V\,d\rho\,\mathbf{B}$. But we may suppose this element to be an element $d\mathfrak{s}$ of volume through which the current $\mathbf{C}$ flows. Thus if for $\gamma\,d\rho$ we write $\mathbf{C}\,d\mathfrak{s}$, and for $\mathbf{F}'$, $\mathbf{F}\,d\mathfrak{s}$, where $\mathbf{F}$ is the force per unit volume exerted by the field, we get

$$\mathbf{F} = V\mathbf{CB}. \tag{61}$$

58. So far we have been able to go by considering the electric field as a mechanical system, but to go further (as Maxwell points out) and find how **B** or **A** depends on the distribution of current and displacement in the field we must appeal to experiment. It has been shewn by experiment that a small circuit produces exactly the same mechanical effects on magnets as would a small magnet, at the same point as the circuit, placed with its positive pole pointing in the direction of the positive normal to the plane of the circuit when the positive direction round the circuit is taken as that of the current*. Moreover the magnetic moment of the magnet which must be placed there is proportional to the strength of the current × the area of the circuit. Further, the effect of this circuit upon other such small circuits is the same as the mutual effects of corresponding magnets. We have now only to consider a finite circuit as split up in the usual way into a number of elementary circuits to see that a finite circuit will act upon magnets or upon other circuits exactly like a magnetic shell of strength proportional to the strength of the current and boundary coinciding with that of the circuit. The unit current in the electro-magnetic system is so taken as to make this proportionality an equality.

The one difference between the circuit and the magnetic shell is that there is no discontinuity in the magnetic potential in going round the circuit, so that by § 50 above the line integral of **H** round the circuit will be $4\pi\times$ the strength of the current. In symbols

$$\int S\mathbf{H}\,d\rho = 4\pi\iint S\mathbf{C}\,d\Sigma$$

for any curve, so that by equation (8) § 6

$$4\pi\mathbf{C} = V\nabla\mathbf{H}, \tag{62}$$

whence

$$S\nabla\mathbf{C} = 0, \tag{63}$$

which of course is a direct result of our original assumption that electricity moves like an incompressible fluid. Maxwell tacitly assumes this by making the assumption that only one coordinate is required to express the motion of electricity in a circuit.

59. We are now in a position to identify the **B** we are now using with the magnetic induction for which we have already used the same symbol.

---

*See § 1 above for the convention with respect to the relation between the positive side of a surface and the positive direction round its boundary. Hitherto we have had no reason for choosing either the right-handed or the left-handed screw as the type of positive and negative rotation. But to make the statement in the text correct we must take the former.

We see by equation (50) § 52 that the mechanical force on the shell corresponding to any coordinate x is

$$-\phi d \iint S \, d\Sigma \, \mathbf{B}'/dx,$$

where $\mathbf{B}'$ is the magnetic induction; and by equations (53) and (55) that the force on the corresponding electric circuit is

$$-\phi d \iint S \, d\Sigma \, \mathbf{B}/dx,$$

therefore $\mathbf{B} = \mathbf{B}'$ wherever there is no magnetism. And where there is magnetism $\mathbf{B}$ is not $= \mathbf{H}$ for $S \nabla \mathbf{B} = 0$, as we have seen. Thus $\mathbf{B} = \mathbf{B}'$ at all points. In other words the two vectors are identical and we are justified in using the same symbol for the two.

This practically ends the general theory of electro-magnetism. We content ourselves with one more application of Quaternions in this subject. We give it because it exhibits in a striking manner the advantages of Quaternion methods.

Electro-magnetic phenomena explained by Stress.

60. In § 46 equation (35) we have seen that the potential energy of a magnetic element $= \mathbf{I} \, d\varsigma$ in a magnetic field is $S \mathbf{IH} \, d\varsigma$ when $\mathbf{H}$ has a potential. Maxwell assumes that the same expression is true whether $\mathbf{H}$ have a potential or not. Assuming this point* with him we can find the force and couple acting on the medium and a stress which will produce that force and couple. The force and couple due to the magnetism of an element is obtained by giving the element an arbitrary translation and rotation and assuming that the work done by this force and couple = the decrement in the potential energy of the element. Thus the force per unit volume is $-\nabla_1 S \mathbf{IH}_1$ for the decrement in the potential energy due to a small translation $\delta\rho$ is $S \, \delta\rho \, \nabla_1 S \mathbf{IH}_1$. Similarly the couple $\mathbf{M}$ is given by

$$\mathbf{M} = V \mathbf{IH} \tag{64}$$

*I do not defend the legitimacy of this assumption. It seems to me bold to assume that a magnet possesses any such thing as potential energy in a field which has no potential. If we assume $\mathbf{H}$ *and its derivatives* to be continuous throughout our typical element $d\varsigma$ of volume containing a great number of molecules (both material and magnetic) the force on a magnetic molecule μ consisting of two poles is $-S \mu \nabla . \mathbf{H}$ and the force per unit volume $-S \mathbf{I} \nabla . \mathbf{H}$, which is only identical with the expression $-\nabla_1 S \mathbf{IH}_1$ obtained below when $\mathbf{H}$ has a potential. With this expression a stress cannot be found that produces the force. If, however, $\mathbf{H}$ and its derivatives be not assumed continuous in this manner the force on the magnet μ is quite indeterminate whether the magnetic pole or the molecular current view of magnetism be taken, unless it be specified in what way the poles and currents are distributed in the element of volume.

for the decrement $-S\mathbf{M}\,\delta\omega$ in the potential energy due to a small rotation $\delta\omega$ is $-S\,V\,\delta\omega\,\mathbf{I}\,.\,\mathbf{H} = -S\;\delta\omega\,V\mathbf{IH}$. The total force $\mathbf{F}$ per unit volume is the sum of that just given and that given by equation (61), so that

$$\mathbf{F} = V\mathbf{CB} - \nabla_1 S\,\mathbf{IH}_1.$$

Therefore by equation (62)

$$\begin{aligned} 4\pi\mathbf{F} &= VV\nabla\mathbf{H}\,.\,\mathbf{B} - 4\pi\nabla_1 S\,\mathbf{IH}_1 \\ &= -\mathbf{H}_1 S\,\nabla_1\mathbf{B} + \nabla_1 S\,\mathbf{BH}_1 - 4\pi\nabla_1 S\,\mathbf{IH}_1 \\ &= -\mathbf{H}_1 S\,\nabla_1\mathbf{B} + \nabla_1 S\,\mathbf{HH}_1. \end{aligned}$$

Now $S\,\nabla\mathbf{B} = 0$ so that $\mathbf{H}_1 S\,\nabla_1\mathbf{B} = \mathbf{H}S\,\Delta\mathbf{B}$, and again

$$\nabla_1 S\,\mathbf{HH}_1 = \nabla(\mathbf{H}^2)/2,$$

so that

$$\mathbf{F} = \phi\Delta, \tag{65}$$

where

$$8\pi\phi\omega = -2\mathbf{H}S\,\omega\mathbf{B} + \omega\mathbf{H}^2. \tag{66}$$

From this we get

$$8\pi V\zeta\phi\zeta = 2V\mathbf{BH} = 8\pi V\mathbf{IH},$$

so that

$$\mathbf{M} = V\zeta\phi\zeta. \tag{67}$$

From these two results (equations (65) and (67)) we see by § 13 above that the stress ϕ will produce all the mechanical effects of the field.

This stress, as can be seen by giving ω the required values in equation (66), is one of pressure $-\mathbf{H}^2/8\pi$ in all directions at right angles to $\mathbf{B}$ and of tension $-S\,\mathbf{H}(2\mathbf{B} - \mathbf{H})/8\pi$ in the direction of $\mathbf{H}$. When there is no magnetism $\mathbf{H} = \mathbf{B}$ so that this pressure and tension become equal and their directions at right angles to and along $\mathbf{B}$ respectively. In fact we then have

$$8\pi\phi\omega = -\mathbf{B}\omega\mathbf{B}. \tag{68}$$

SECTION V.

HYDRODYNAMICS.

61. In the applications I am about to make in this I have practically nothing new to shew except the utility of Quaternion methods in the general theory of Hydrodynamics in all its parts.

I therefore take a treatise (Greenhill's article in the *Encyc. Brit.* on this subject) and work out the general theory on the lines of the treatise. This is more necessary than at first sight it would seem, for I believe mathematicians who have studied Quaternions are under the impression that the method does not lend itself conveniently to the establishment and treatment of such equations as the Lagrangian and those of Cauchy's integrals. With our meaning of ∇, however, the Quaternion treatment of these equations is as much simpler than the Cartesian as in the case of the Eulerian equations.

Down to equation (13) below the subject has already been handled by Prof. Hicks* in his *Quaternion treatment of Strains and Fluid motion* (*Quart. Journ. Math.* XIV. [1877] p. 271). I do not hesitate to go over the ground again as my methods are different from his.

Notation.

62. For the vector velocity at any point we shall use σ, for the density D, for the force per unit mass **F**, for its potential, when it has one, v, and for the pressure p. For time-flux which follows a particle we shall use d/dt or the Newtonian dot, and for that which refers to a fixed point of space $\partial/\partial t$†.

Thus

$$d/dt = \partial/\partial t - S\sigma\nabla. \qquad (1)$$

*[Note added, 1892. Prof. Tait's name ought to be added to Prof. Hicks's.]

†[Note added, 1892. I am aware that this is contrary to the usual English custom, but that custom—of interchanging the meanings of d/dt and $\partial/\partial t$ as given in the text—seems to me out of harmony with the meaning attached to ∂ in other branches of Mathematics. At any rate I have respectable fellow-sinners, e.g. Kirchhoff in his *Mechanik*, zweite Vorlesung, et seq.]

This equation is given on p. 446 of Greenhill's article already mentioned. In future we shall refer to this article simply as "Greenhill's article."

Euler's Equations.

63. To find the equation of continuity, with Greenhill, we merely express symbolically that the rate of increase of the mass of the fluid in any space equals the rate at which it is flowing through the boundary. Thus M being the mass in any space,

$$\partial M/\partial t = \iint DS\sigma\, d\Sigma. \tag{2}$$

This is Greenhill's equation (1) p. 445. By it and equation (9) § 6 above we have

$$\partial M/\partial t = \iiint S\nabla(D\sigma)\, d\mathfrak{s},$$

whence reducing the volume to the element $d\mathfrak{s}$,

$$\partial D/\partial t = S\nabla(D\sigma). \tag{3}$$

This is Greenhill's equation (2). Thus by equation (1) of last section,

$$\dot{D}/D = S\nabla\sigma. \tag{4}$$

64. To obtain Euler's equations of motion we express that the vector sum of the impressed forces for any volume equals the vector sum of the bodily forces plus the vector sum of the pressures on the surface. Thus

$$\begin{aligned}\iiint D\dot{\sigma}\, d\mathfrak{s} &= \iiint D\mathbf{F}\, d\mathfrak{s} - \iint p\, d\Sigma \\ &= \iiint (D\mathbf{F} - \nabla p)\, d\mathfrak{s},\end{aligned}$$

by equation (9) § 6 above. Applying this to the element $d\mathfrak{s}$ and dividing by $D\,d\mathfrak{s}$ we have

$$\dot{\sigma} = \mathbf{F} - \nabla p/D, \tag{5}$$

or by equation (1) § 62

$$\partial\sigma/\partial t - S\sigma\nabla\,.\,\sigma = \mathbf{F} - \nabla p/D. \tag{6}$$

This is Greenhill's equations (4) (5) (6).

If $F(\rho, t) = 0$ (F a scalar) be the equation of a surface always containing the same particles $\dot{F} = 0$, or by equation (1)

$$\partial F/\partial t - S\sigma\nabla F = 0. \tag{7}$$

This is Greenhill's equation (7).

65. Let us now put

$$\int dp/D = P. \tag{8}$$

This of course assumes that D is a function of p only, which is not always the case, for instance in a gas where diffusion of heat is taking place. If **F** have a potential v, $\mathbf{F} = -\nabla v$. Thus equation (5) becomes

$$\dot{\sigma} = -\nabla(v + P), \tag{9}$$

and equation (6)

$$\partial\sigma/\partial t - S\sigma\nabla \,.\, \sigma = -\nabla(v + P). \tag{10}$$

Now

$$S\sigma\nabla_1 \,.\, \sigma_1 = V\sigma V\nabla_1\sigma_1 + \nabla_1 S\sigma\sigma_1 = V\sigma V\nabla\sigma + \nabla(\sigma^2)/2.$$

Thus equation (10) becomes

$$\partial\sigma/\partial t + 2V\epsilon\sigma + \nabla R = 0, \tag{11}$$

where the scalar R is put for $P + v - \sigma^2/2$ and the vector ϵ for $V\nabla\sigma/2$. This is Greenhill's equations (8) (9) (10).

If $\epsilon = 0$, i.e. $V\nabla\sigma = 0$, we have $\sigma = \nabla\phi$, whence our equation becomes

$$\nabla(\partial\phi/\partial t + R) = 0,$$

so that

$$\partial\phi/\partial t + R = H, \tag{12}$$

where H is a function of t only. In next section we shall obtain in the case of an infinite fluid a generalisation of this which I believe has not hitherto been obtained. Here we have made the assumption that if $V\nabla\sigma = 0$ at one epoch it will be so always. This we shall prove later.

66. Greenhill next considers the case of steady motion. In this case $\partial\sigma/\partial t = 0$, so that equation (11) becomes

$$2V\epsilon\sigma + \nabla R = 0, \tag{13}$$

and therefore the surface $R =$ const. contains both vortices and stream-lines and the relation $dR/dn = 2q\omega \sin\theta$ given on p. 446 of Greenhill's article is the natural interpretation of our equation.

So far we have been going over much the same ground as Hicks, but now we enter upon applications of Quaternions that I think have not been made before.

67. Greenhill next considers rotating axes and finds the form of the equations of motion when referred to these. Let σ be the velocity referred to them; so that if the axes have at any time made the rotation $q(\ )q^{-1}$ the real velocity will be $q\sigma q^{-1}$. Thus*, as always with rotating axes, if α be any vector function of a particle the rate of increase of α in space is

$$\dot{\alpha} + V\omega\alpha,$$

where ω is the angular velocity referred to the same system of rotating axes. Thus we see that the acceleration of a particle $= \dot{\sigma} + V\omega\sigma$. The velocity $\sigma = \dot{\rho}' + V\omega\rho'$, where ρ' is the vector coordinate referred to our present axes (i.e. in the notation of the footnote $\rho' = q\rho q^{-1}$). Thus our equation of motion is now

$$\dot{\sigma} + V\omega\sigma = \mathbf{F} - {}_{\rho'}\nabla p/D,$$

but now

$$d/dt = \partial/\partial t - S\dot{\rho}'{}_{\rho'}\nabla. = \partial/\partial t - S(\sigma - \omega\rho')_{\rho'}\nabla,$$

whence changing ρ' into ρ we have

$$\partial\sigma/\partial t - S(\sigma - \omega\rho)\nabla \,.\, \sigma + V\omega\sigma = \mathbf{F} - \nabla p/D, \qquad (14)$$

which is the equation Greenhill gives on p. 446.

*This is quoted as a known result because it occurs generally in the subject of *Rigid Dynamics*. No Quaternion proof, however as far as I am aware, has been given. We therefore give one here. What is meant by rotating axes may be thus explained.—Instead of choosing as our coordinates the vectors $\alpha', \beta' \ldots$ which occur in any problem, we take others $\alpha, \beta \ldots$ such that $\alpha' = q\alpha q^{-1}$, $\beta' = q\beta q^{-1} \ldots$ $q(\ )q^{-1}$ may be called the integral rotation of the axes. Thus if we say that the vector angular velocity of these axes themselves is ω we mean that the real angular velocity is $q\omega q^{-1}$, so that, as can be seen by putting in equation (a) below α = const., or as in equation (58) § 33 above, $\omega = 2Vq^{-1}\dot{q}$. (This maybe established also by Tait's *Quaternions*, § 356, equation (2), from which $q\omega q^{-1} = 2V\dot{q}q^{-1}$ or $\omega = 2Vq^{-1}\dot{q}$). Again, when we say that the rate of increase of α in space is τ, we mean that $\dot{\alpha}' = q\tau q^{-1}$ or $\tau = q^{-1} \,.\, d(q\alpha q^{-1})/dt \,.\, q$, or

$$\tau = \dot{\alpha} + 2VVq^{-1}\dot{q} \,.\, \alpha, \qquad (a)$$

or

$$\tau = \dot{\alpha} + V\omega\alpha. \qquad (b)$$

This could have been proved with fewer symbols and more explanation, but the above seems to me the most characteristic Quaternion proof. We might have started with not quite so general an explanation of reference to rotating axes and so refrained from introducing the *integral* rotation, and therefore also q.

The Lagrangian Equations.*

68. We now consider the history of a single particle and for this we require different notation.

We consider the vector coordinate (ρ) of a particle as a function of some other vector α (say the initial value of ρ) and of t.

We first require the connection between ${}_\alpha\nabla$ and ${}_\rho\nabla$. We shall drop the affix α and retain ρ, so that now *not* $Q(\nabla_1, \rho_1)$ but $Q(\nabla_1, \alpha_1)$

$$= Q(\zeta, \zeta).$$

Now $$S\, d\alpha\, \nabla = S\, d\rho\, {}_\rho\nabla,$$

but $$d\rho = -\rho_1 S\, d\alpha\, \nabla_1,$$

$$\therefore \quad S\, d\alpha\, \nabla = -S\, d\alpha\, \nabla_1 S \rho_1 {}_\rho\nabla,$$

or since $d\alpha$ is perfectly arbitrary

$$\nabla = -\nabla_1 S \rho_1 {}_\rho\nabla. \tag{15}$$

Thus operating upon equation (9) § 65 by $\nabla_1 S \rho_1(\ )$ and remembering that in that equation ∇ must be changed to ${}_\rho\nabla$ we get

$$\nabla_1 S \rho_1 \dot{\sigma} = \nabla(v + P), \tag{16}$$

and this is our new equation of motion (Greenhill's (1) (2) (3) p. 448).

In equation (1) § 12 above let us put for $\rho + \eta$, ρ; and for ρ, α.

Thus $$\chi\omega = -S\omega\nabla \,.\, \rho. \tag{17}$$

Hence by equation (6*d*) § 14 above,

$$\text{strained vol. of el./original vol.} = S\nabla_1\nabla_2\nabla_3 S\rho_1\rho_2\rho_3/6,$$

whence we see that

$$DS\nabla_1\nabla_2\nabla_3 S\rho_1\rho_2\rho_3 = 6D_0, \tag{18}$$

where D_0 is a constant which when α is taken as the initial value of ρ is the original density. This is the equation of continuity.

---

*[Note added, 1892. At the time of writing the essay I did not notice that these equations are a particular case of the general equation for an elastic body already established (see equations (15*m*) § 16 and (15*n*) § 16*a*)

$$D\ddot{\rho}' = \mathbf{F} - 2\rho_1' S\nabla_1 \Psi ꟼ w\, \Delta.]$$

Cauchy's integrals of these equations.

69. To obtain these integrals we require to express ${}_\rho\nabla$ in terms of ∇. Now equation (15) of last section expresses ∇ as a linear function of ${}_\rho\nabla$. The converse we have already seen how to get. In fact from equation (6*h*) § 3*a* above

$$2J{}_\rho\nabla = -V\rho_1\rho_2\, S\nabla_1\nabla_2\nabla, \tag{19}$$

where, with Greenhill, for brevity J is put for $S\nabla_1\nabla_2\nabla_3\, S\rho_1\rho_2\rho_3/6$. The only function we wish to apply this to is $V{}_\rho\nabla\sigma$. We have

$$\begin{aligned} 2JV{}_\rho\nabla\sigma &= V\sigma_3\, V\rho_1\rho_2\, S\nabla_1\nabla_2\nabla_3 \\ &= (\rho_2 S\rho_1\sigma_3 - \rho_1 S\rho_2\sigma_3)\, S\nabla_1\nabla_2\nabla_3, \end{aligned}$$

or interchanging the suffixes 1 and 2 in the last term

$$JV{}_\rho\nabla\sigma = \rho_2 S\rho_1\sigma_3\, S\nabla_1\nabla_2\nabla_3, \tag{20}$$

which gives the spin at any instant in terms of our present independent variables α and t.

70. In order to obtain Cauchy's integral of equation (16) operate on it by $V\nabla(\ )$. Thus

$$0 = V(\nabla_1 + \nabla_2)\nabla_1 S\rho_1\dot{\sigma}_2 = S\rho_1\dot{\sigma}_2\,.\,V\nabla_2\nabla_1.$$

Now

$$d(S\rho_1\sigma_2\,.\,V\nabla_2\nabla_1)/dt = S\sigma_1\sigma_2\,.\,V\nabla_2\nabla_1 + S\rho_1\dot{\sigma}_2\,.\,V\nabla_2\nabla_1.$$

The first term of this last expression is zero since the sign is changed by interchanging the suffixes. From the last expression then

$$\begin{aligned} S\rho_1\sigma_2\,.\,V\nabla_2\nabla_1 &= \text{constant} \\ &= \text{initial value of } S\zeta\sigma_2 V\nabla_2\zeta = -V\nabla\sigma_0, \end{aligned}$$

where σ_0 is the initial value of σ. Changing the suffixes and substituting in equation (20) we have

$$JV{}_\rho\nabla\sigma = -\rho_1 S\nabla\sigma_0\nabla_1,$$

or giving J its value D_0/D from equation (18) and putting $V{}_\rho\nabla\sigma = \epsilon$, $V\nabla\sigma_0 = \epsilon_0$, we have

$$\epsilon/D = -S(\epsilon_0/D_0)\nabla\,.\,\rho. \tag{21}$$

This is Greenhill's equations (4) (5) (6) p. 448.

The physical interpretation of this equation is quite easy. Consider a small vector $d\alpha$ drawn in the fluid initially. At the time t this will have become $d\rho = -S\, d\alpha\nabla\,.\,\rho$. Thus we see that if a small vector $c\epsilon_0/D_0$ be drawn in the fluid initially it will at the time t be $c\epsilon/D$, from which we infer that an element of a vortex filament will always remain an element of a vortex filament; or, a vortex filament or tube always remains a vortex filament or tube. Again we see that $T\epsilon/D$ at any time varies directly with the elongation in the direction of the vortex filament so that $T\epsilon$ varies as that elongation $\times$ the density, i.e. inversely as the cross-section of a small vortex tube at the point. This is not the easiest way of arriving at these results, but it is well to show in passing how easy of interpretation are our results.

We see from equation (21) that if $\epsilon_0 = 0$, $\epsilon = 0$. In other words, if the motion have a velocity potential at one instant it will have one always.

Flow, circulation, vortex-motion.

71. We are about to consider vortex-motion from another point of view, viz. that of circulation.

In § 12 we saw that a strain due to a small displacement η could be decomposed into a pure strain followed by a rotation, the vector rotation being $V\nabla\eta/2$. If now for η we put the small vector $\sigma\,\delta t$ we see the propriety of calling $V\nabla\sigma/2$ the *spin*. This therefore is taken as a *definition* of spin. Greenhill does not take this (usual) course but uses the property we shall immediately prove concerning circulation and spin to lead to his definition.

It is not necessary here to define flow and circulation. Putting $\sigma = \nabla\phi$ we see that for irrotational motion the flow $= -\int S\sigma d\rho = \int d\phi$ from one point to another is the increment in the velocity potential. Thus for mutually reconcilable paths it is always the same.

Taking the circulation round a closed curve in the general case, the curve not inclosing any singular region of the fluid, we may transform the line integral into a surface integral by equation (8) § 6 above. Thus

$$-\int S\sigma\, d\rho = -\iint S\, d\Sigma\,\nabla\sigma, \tag{22}$$

so that the circulation round the curve equals twice the surface integral of the spin. Hence Greenhill's definition of the spin.

72. From equation (9) § 65 we see that

$$-\frac{d}{dt}\left(\int_A^B S\sigma\, d\rho\right) = -\int_A^B S\sigma\, d\sigma - \int_A^B S\dot{\sigma}\, d\rho$$
$$= -\tfrac{1}{2}\int_A^B d(\sigma^2) - \int_A^B d(v+P),$$

or
$$-\frac{d}{dt}\left(\int_A^B S\sigma\, d\rho\right) = -\left[v + P + \sigma^2/2\right]_A^B. \qquad (23)$$

Therefore for a closed curve

$$d(\textstyle\int S\sigma\, d\rho)/dt = 0,$$

so that the circulation round the curve, and therefore the corresponding surface integral of the spin remains always constant. Taking the curve round a small vortex tube we once more arrive at the propositions enunciated in § 70 about vortices.

Thus at any point of a vortex tube the strength which is defined as the product of the spin into the cross-section is constant throughout all time. Also it is the same for all points of the tube, for by equation (9) § 6 we have

$$\iint S\, d\Sigma\, \nabla\sigma = \iiint S\,\nabla^2\sigma\, d\mathfrak{s} = 0$$

for any portion of the tube. But the only part contributing to the surface integral is the ends of the part of the tube considered, so that the strength at these two ends is the same, and is therefore constant for the whole tube.

73. These propositions are proved in the paper by Hicks already referred to. There is yet a third method of proof which, like Hicks's, is derived directly from the equation (9) of motion.

We have from equation (1) § 62

$$\frac{d}{dt}\nabla - \nabla\frac{d}{dt} = -S\sigma\nabla\,.\,\nabla + \nabla S\sigma\nabla. = \nabla_1 S\sigma_1\nabla.\,; \qquad (24)^*$$

therefore $\quad dV\nabla\sigma/dt - V\nabla\dot{\sigma} = V(\nabla_1 S\sigma_1\nabla_2\,.\,\sigma_2).$

*[Note added, 1892. When first giving this in the *Mess. of Math.* 1884 and when again putting it in the present essay, I was unaware of Prof. Tait's paper in *Proc. R. S. E.* 1869–70, p. 143, where in the present notation he has for an incompressible fluid, (1) $d/dt = \partial/\partial t - S\sigma\nabla$ [given

Now by equation (9) § 65, $V\nabla\dot{\sigma} = 0$, so that

$$dV\nabla\sigma/dt = V\nabla_1\sigma_2 S\nabla_2\sigma_1, \tag{25}$$

$$\therefore \qquad d(\rho + xV\nabla\sigma)/dt = \sigma + \dot{x}V\nabla\sigma + xV\nabla_1\sigma_2 S\nabla_2\sigma_1.$$

Now $\sigma_1 S\,\nabla_2\sigma_2\nabla_1 = V\sigma_2\nabla_1 S\,\nabla_2\sigma_1 + V\nabla_1\nabla_2 S\,\sigma_1\sigma_2 + V\nabla_2\sigma_2 S\,\nabla_1\sigma_1$, but $V\nabla_1\nabla_2 S\,\sigma_1\sigma_2 = 0$, for an interchange of suffixes leads to a change of sign. Therefore

$$V\nabla_1\sigma_2 S\,\nabla_2\sigma_1 = V\nabla\sigma S\,\nabla\sigma - S\,\nabla\sigma\nabla\,.\,\sigma.$$

Put now $x = c/D$ where c is some small constant. Thus

$$\dot{x}/c = -\dot{D}/D^2 = -S\,\nabla\sigma/D,$$

and we get
$$d(\rho + cV\nabla\sigma\,.\,/D)\,dt = \sigma - cS\,\nabla\sigma\,\nabla\,.\,\sigma\,.\,/D. \tag{26}$$

This shews that $\rho' \equiv \rho + cV\nabla\sigma/D$ is the variable vector of a particle, for the equation asserts that the time-flux of ρ' = the velocity at ρ'. Thus again we get the laws of vortices given in § 70.

Irrotational Motion.

74. We use this heading merely to connect what follows with what Greenhill has under the same on p. 449. It is not very appropriate here.

For irrotational motion we have seen that we may put

$$\sigma = \nabla\phi. \tag{27}$$

If the fluid be incompressible we have further

$$0 = S\,\nabla\sigma = \nabla^2\phi. \tag{28}$$

Let T be the kinetic energy of this liquid. Thus

$$-2T = D\iiint(\nabla\phi)^2\,ds = D\iint\phi S\,d\Sigma\,\nabla\phi - D\iiint\phi\nabla^2\phi\,ds$$

in Cartesian form], (2) $\dot{\sigma} = -\nabla v - \nabla p/D$, (3) $S\,\nabla\sigma = 0$, (4) $\nabla\dot{\sigma} - d(\nabla\sigma)/dt = -S\,\nabla\sigma\nabla\,.\,\sigma$, (5) $V\nabla\dot{\sigma} = 0$, (6) $d\nabla\sigma/dt = -S\,\nabla\sigma\nabla\,.\,\sigma$. Perhaps I have not interpreted Prof. Tait's notation which is but briefly described correctly, but (4) should apparently be $V\nabla\dot{\sigma} - d(\nabla\sigma)/dt = S\,\nabla\sigma\nabla\,.\,\sigma$, and so agree with (5) and (6). It will be seen that the whole of §§ 512–13 of Tait's *Quaternions*, 3rd ed., is contained in this essay. I cannot at present recall whether this is owing wholly or in part to my being indebted to other old papers of Prof. Tait, or whether in writing his 3rd ed. he has arrived independently at the same treatment, but am inclined to the latter belief.]

by equation (9) § 6 above. Thus by equation (28)

$$T = -\tfrac{1}{2}D\iint \phi S\, d\Sigma\, \nabla\phi. \tag{29}$$

Hence we see that if we have a vector σ which satisfies the equations $V\nabla\sigma = 0$ and $S\nabla\sigma = 0$ throughout any singly-connected space, or more generally satisfies the second of these equations throughout *any* space, and also has zero circulation round any closed curve in that space*; and further satisfies the equation $S\, d\Sigma\, \sigma = 0$ at the boundary; then the function

$$\iiint \sigma^2\, d\mathfrak{s} = 0,$$

or $\sigma = 0$ at all points of the space, for each element of the integral being essentially negative must be zero.

75. The following important theorem now follows: If there are given; at every point of any region the convergence $S\nabla\sigma$ and the spin $V\nabla\sigma$, at every point of the boundary the normal velocity (and therefore $S\, d\Sigma\, \sigma$), and the circulation round every cavity which increases the cyclomatic number of the region; then the motion (as given by σ at every point) if possible, is unique. For let σ be one possible velocity, and if possible $\sigma + \tau$ another. Thus τ satisfies all the conditions that σ does at the end of last section, and therefore $\tau = 0$ at every point or σ is unique.

Motion of a solid through a liquid.

76. We assume that there is no circulation of the liquid for any cycle; in other words, that if all the solids be brought to rest, so will be also the liquid.

We shall take axes of reference fixed in the (one) moving solid. In the footnote to § 67 is explained how the *rotation* of these axes is taken account of. The effect of the *translation* of the origin will be easily found.

Let σ, ω be the linear and angular velocities respectively of the moving solid. Let us put

$$\phi = -S\sigma\psi - S\omega\chi, \tag{30}$$

where ϕ is the required velocity potential of the liquid, and ψ and χ are two vector functions of the position of a point independent of σ and ω. Let us see whether

*To insure that ϕ is single valued and so, that the part of the surface integral of equation (29) due to "barriers" is zero.

ψ and χ *can* be found so as to satisfy these conditions. σ and ω are of course assumed as quite arbitrary.

The conditions are first the equation (28) of continuity

$$\nabla^2\phi = 0, \tag{28}$$

which gives

$$\nabla^2\psi = 0, \nabla^2\chi = 0, \tag{31}$$

and second, the equality of the normal velocity $-S\,U\,d\Sigma\,\nabla\phi$ of the liquid at any point of the boundary with that of the boundary at the same point. Thus for a fixed boundary

$$S\sigma(S\,d\Sigma\,\nabla\,.\,\psi) + S\omega(S\,d\Sigma\,\nabla\,.\,\chi) = 0,$$

whence on account of the arbitrariness of σ and ω

$$S\,d\Sigma\,\nabla\,.\,\psi = S\,d\Sigma\,\nabla\,.\,\chi = 0. \tag{32}$$

For a moving boundary again we have

$$S\,d\Sigma\,\nabla\phi = S\,d\Sigma\,(\sigma + V\omega\rho),$$

or

$$S\sigma(d\Sigma + S\,d\Sigma\,\nabla\,.\,\psi) + S\omega(V\rho\,d\Sigma + S\,d\Sigma\,\nabla\,.\,\chi) = 0,$$

which gives

$$d\Sigma + S\,d\Sigma\,\nabla\,.\,\psi = 0, \tag{33}$$

$$V\rho\,d\Sigma + S\,d\Sigma\,\nabla\,.\,\chi = 0. \tag{34}$$

Now it is well known that ψ and χ can be determined to satisfy all these conditions. In fact x being a coordinate of either ψ or χ these conditions amount to:— $\nabla^2 x = 0$ throughout the space, and $S\,d\Sigma\,\nabla x =$ given value at the boundary.

77. We do not propose to find ψ and χ in any particular case. We leave p. 454 of Greenhill's article and go on to p. 455, i.e. we proceed to find the equations of motion of the solid. For this purpose we require the kinetic energy of the system. Calling this T we shall have

$$T = -S\sigma\,\Sigma\sigma/2 - S\omega\,\Phi\sigma - S\omega\,\Omega\omega/2, \tag{35}$$

where Σ, Ω and Φ are linear vector functions and the first two are self-conjugate. This is the most general* quadratic function of ω and σ. It involves 21 independent constants, six in Σ, six in Ω and nine in Φ. When ψ and χ are known Σ, Φ and Ω are all known. We will obtain the expressions for them, and for simplicity will assume that the origin is at the centre of gravity of the moving solid. Let M be the mass of this last and $\mu\omega$ (where, as is well-known, μ is a self-conjugate linear vector function) the moment of momentum. Thus by equation (29)

$$2T = -D\iint \phi S\, d\Sigma\, \nabla\phi - M\sigma^2 - S\,\omega\mu\omega.$$

Putting $S\, d\Sigma\, \nabla\phi = S\, d\Sigma(\sigma + V\omega\rho)$ at the moving boundary and zero at the fixed,

$$2T = D\iint (S\,\sigma\psi + S\,\omega\chi)(S\,\sigma\, d\Sigma + S\,\omega\rho\, d\Sigma) - M\sigma^2 - S\,\omega\,\mu\omega$$

where the surface integral must be taken only over the moving boundary. Thus noting that Σ and Ω are self-conjugate

$$\Sigma\lambda = M\lambda - \tfrac{1}{2}D\iint(\psi S\,\lambda\, d\Sigma + d\Sigma\, S\,\lambda\psi), \tag{36}$$

$$\Omega\lambda = \mu\lambda - \tfrac{1}{2}D\iint(\chi S\,\lambda\rho\, d\Sigma + V\rho\, d\Sigma\, S\,\lambda\chi), \tag{37}$$

$$-S\,\lambda\Phi\lambda' = \tfrac{1}{2}D\iint(S\,\lambda\chi\, S\,\lambda'\, d\Sigma + S\,\lambda\rho\, d\Sigma\, S\,\lambda'\psi). \tag{38}$$

These surface integrals can be simplified, for in each of these equations the first surface integral equals the second. In equation (36) in order to prove this we have merely to put for $d\Sigma$, $-S\, d\Sigma\, \nabla\,.\,\psi$ (equation (33)) when we shall find by equation (9) § 6 above (since we may now suppose the integrations to extend over the *whole* boundary), that

$$\iint \psi S\,\lambda\, d\Sigma = -\iiint \psi_1 S\,\nabla_1\nabla_2 S\,\lambda\psi_2\, ds = \iint d\Sigma\, S\,\lambda\psi.$$

Similarly for equation (37). Again in equation (38),

$$\iint S\,\lambda\chi\, S\,\lambda'\, d\Sigma = -\iiint S\,\nabla_1\nabla_2\, S\,\lambda\chi_1\, S\,\lambda'\psi_2\, ds = \iint S\,\lambda\rho\, d\Sigma\, S\,\lambda'\psi.$$

Thus finally for Σ, Ω and Φ

$$\Sigma\lambda = M\lambda - D\iint \psi S\,\lambda\, d\Sigma = M\lambda - D\iint d\Sigma\, S\,\lambda\psi, \tag{39}$$

$$\Omega\lambda = \mu\lambda - D\iint \chi S\,\lambda\rho\, d\Sigma = \mu\lambda - D\iint V\rho\, d\Sigma\, S\,\lambda\chi, \tag{40}$$

$$-S\,\lambda\Phi\lambda' = D\iint S\,\lambda\chi\, S\,\lambda'\, d\Sigma = D\iint S\,\lambda\rho\, d\Sigma\, S\,\lambda'\psi. \tag{41}$$

*For let $A(\sigma,\sigma)+B(\omega,\sigma)+C(\omega,\omega)$ be this general function A, B, and C being scalar functions each linear in each of its constituents, and let

$$\Sigma\sigma = \zeta\{A(\zeta,\sigma) + A(\sigma,\zeta)\}, \quad \Phi\sigma = \zeta B(\zeta,\sigma), \quad \Omega\omega = \zeta\{C(\zeta,\omega) + C(\omega,\zeta)\}.$$

This last may be put in the following four forms

$$\Phi'\lambda = -D\iint d\Sigma\, S\,\lambda\chi = -D\iint \psi S\,\lambda\rho\, d\Sigma, \tag{42}$$

$$\Phi\lambda = -D\iint \chi S\,\lambda\, d\Sigma = -D\iint V\rho\, d\Sigma\, S\,\lambda\psi. \tag{43}$$

Thus assuming that ψ and χ are determined we have found T as a quadratic function of σ and ω.

78. If **P**, **G** be the linear and angular impulses of the system respectively our equations of motion are

$$\dot{\mathbf{P}} + V\omega\mathbf{P} = \mathbf{F},$$

$$\dot{\mathbf{G}} + V\omega\mathbf{G} = \mathbf{M},$$

by the footnote to § 67 above. Here **F** and **M** are the external force and couple applied to the body. Now at the instant under consideration $\mathbf{P} = {}_\sigma\nabla T$, $\mathbf{G} = {}_\omega\nabla T$. But if the origin had been at the point $-\rho$, **P** would still be ${}_\sigma\nabla T$ whereas **G** would be ${}_\omega\nabla T + V\rho\mathbf{P}$. Thus differentiation with regard to t does not affect the form of **P** but does that of **G**. In fact using the last stated values of **P** and **G** along with the last two equations and eventually putting $\dot{\rho} = \sigma$, $\rho = 0$ we get

$$d_\sigma\nabla T/dt + V\omega_\sigma\nabla T = \mathbf{F}, \tag{44}$$

$$d_\omega\nabla T/dt + V\omega_\omega\nabla T + V\sigma_\sigma\nabla T = \mathbf{M}. \tag{45}$$

Now from equation (35) we see that

$${}_\sigma\nabla T = \Sigma\sigma + \Phi'\omega, \tag{46}$$

$${}_\omega\nabla T = \Phi\sigma + \Omega\omega, \tag{47}$$

whence from equations (44) and (45)

$$\Sigma\dot{\sigma} + \Phi'\dot{\omega} + V\omega\,\Sigma\sigma + V\omega\,\Phi'\omega = \mathbf{F}, \tag{48}$$

$$\Phi\dot{\sigma} + \Omega\dot{\omega} + V\sigma\,\Sigma\sigma + (V\sigma\,\Phi'\omega + V\omega\,\Phi\sigma) + V\omega\,\Omega\omega = \mathbf{M}. \tag{49}$$

Thus we see that with Quaternion notation even the general equations of motion are not too complicated to write down conveniently.

We now leave Greenhill's article and proceed to certain theorems not contained therein. The Cartesian treatment of these subjects will be found in Lamb's treatise on *The Motion of Fluids*, Chapters VI. and IX., which are headed *Vortex Motion* and *Viscosity* respectively.

The velocity in terms of the convergences and spins.

79. We have seen in § 75 that if the spin and convergence are given for each point of a bounded fluid and the normal velocity at each point of the boundary, there is if any, but one possible motion. We shall see directly that this unique value always exists.

At present we observe that we cannot find a motion giving an assigned spin and convergence for each point and *any* assigned velocity for each point of the boundary. If however with such data a motion be possible we can find the velocity at any point explicitly. By equation (19) § 10 above we have

$$4\pi\sigma = -\nabla \iiint \nabla u\sigma \, d\mathfrak{s},$$

but by equation (9) § 6

$$\iiint \nabla u\sigma \, d\mathfrak{s} = \iint u \, d\Sigma \, \sigma - \iiint u \nabla\sigma \, d\mathfrak{s}$$

$$\therefore \quad 4\pi\sigma = \iint [V]\nabla u \, d\Sigma \, \sigma - \iiint [V]\nabla u \, \nabla\sigma \, d\mathfrak{s}, \tag{50}$$

where the square brackets indicate that we may or may not retain the V at our convenience. This equation may be put

$$4\pi\sigma = \iint [V](\nabla u \, d\Sigma \, \sigma - u \, d\Sigma \, \nabla\sigma) + \iiint u \nabla^2 \sigma \, d\mathfrak{s}. \tag{51}$$

Both of these equations solve the question now proposed.

If the fluid be considered infinite the surface integral of equation (50) vanishes if σ converges to zero at infinity and also that of equation (51) if in addition the spin and convergence converge to a quantity infinitely small compared with the reciprocal of the distance of the surface.

80. We may now consider the case when $\nabla\sigma$ (spin and convergence) is given at all points and $S \, d\Sigma \, \sigma$ (normal velocity) at the boundary. It must be observed that the given value of the spin must be distributed in a solenoidal manner for

$$S \nabla V \nabla\sigma = 0.$$

Whenever the quaternion

$$4\pi q = -\iiint \nabla u \, \nabla\sigma \, d\mathfrak{s} = \nabla \iiint u \, \nabla\sigma \, d\mathfrak{s}, \tag{52}$$

is a vector, we see that all the conditions except the boundary ones are satisfied by putting $\sigma = q$. Let us then see when q reduces to a vector. By equation (9) § 6 above

$$4\pi S q = -\iint uS \, d\Sigma \, \nabla\sigma + \iiint uS \, \nabla^2\sigma \, d\mathfrak{s}.$$

Thus q is a vector if this surface integral vanish. The surface integral vanishes if all the vortices form closed curves within the space. If they do not we must extend the space and make them form closed curves outside the original space. Extending the volume integral accordingly, we may put

$$4\pi\sigma' = -\iiint \nabla u \, \nabla\sigma \, d\mathfrak{s}, \tag{53}$$

and now $\nabla\sigma' = \nabla\sigma$. Suppose then that

$$\sigma = \sigma' + \sigma''. \tag{54}$$

We thus get $\nabla\sigma'' = 0$ and may therefore put

$$\sigma'' = \nabla\phi, \tag{56}$$

where ϕ satisfies the equations

$$\nabla^2\phi = 0, \tag{57}$$

and $$S\, d\Sigma \, \nabla\phi = S\, d\Sigma(\sigma - \sigma') = \text{known quantity.} \tag{58}$$

Now it is well known that ϕ can be determined so as to satisfy these two equations. Therefore the problem under discussion always admits of solution.

The above is equivalent to Lamb's §§ 128–131. His § 130 is the natural interpretation of the equation

$$4\pi\sigma = -\iiint V\nabla u \, \nabla\sigma \, d\mathfrak{s}.$$

His § 132 is seen at once from equation (50) above. For in the case he considers we, in accordance with § 7 above, take each side of the surface of discontinuity as a part of the boundary. Now $[S\, d\Sigma\, \sigma]_{a+b} = 0$ so that for this part of the boundary we can leave out the part $S\, d\Sigma\, \sigma$ and we get

$$4\pi\sigma = \iint V\nabla u \, V\, d\Sigma\, \sigma - \iiint V\nabla u \, \nabla\sigma \, d\mathfrak{s}, \tag{59}$$

so that if we regard $[V\, d\Sigma\, \sigma]_{a+b}$ as $-2\times$ an element of a vortex, we get the same law for these vortex *sheets* as for the vortices in the rest of the fluid.

81. The velocity potential due to a single vortex filament of strength $d\theta$ is at once obtained by putting in equation (50) for $V\nabla\sigma\, d\mathfrak{s}$, $2\, d\theta\, d\rho$. Thus calling σ' the part of the velocity due to the filament

$$2\pi\sigma' = -d\theta \int V\nabla u \, d\rho = -d\theta \iint u_1 V\nabla_1 V\, d\Sigma\, \nabla_1$$

by equation (8) § 6 above. Now

$$V\nabla_1 V\, d\Sigma\, \nabla_1 = -d\Sigma\, \nabla_1^2 + \nabla_1 S\, d\Sigma\, \nabla_1.$$

Hence, since $\nabla^2 u = 0$ for all points not on the filament

$$2\pi\sigma' = \nabla(d\theta \iint S\, d\Sigma\, \nabla u). \tag{60}$$

Thus the velocity potential $= (2\pi)^{-l}\times$ the strength $\times$ the solid angle subtended by the filament at the point considered.

Kinetic Energy.

82. Let us put

$$4\pi q = \iiint u\nabla\sigma\, d\mathfrak{s}, \tag{61}$$

so that

$$\sigma = \nabla q. \tag{62}$$

We may now find expressions for T, the kinetic energy, in one or two interesting forms. We have

$$2T = -\iiint D\sigma^2\, d\mathfrak{s},$$

whence

$$2T = -\iiint DS\,\sigma\,\nabla q\, d\mathfrak{s} = -\iint DS\,\sigma\, d\Sigma\, q + \iiint D_1 S\,\sigma_1\, \nabla_1 q\, d\mathfrak{s}.$$

We assume that the fluid extends to infinity and that the vortices and convergences are all at a finite distance, so that at infinity σ is of order $1/R^2$ and q of order $1/R$. Thus the surface integral vanishes and we have

$$2T = \iiint D_1 S\,\sigma_1\, \nabla_1 q\, d\mathfrak{s}, \tag{63}$$

or, putting in the value of q from equation (61),

$$T = (8\pi)^{-1} \iiint\iiint u D_1 S\,\sigma_1\, \nabla_1 \nabla_2 \sigma_2\, d\mathfrak{s}_1\, d\mathfrak{s}_2. \tag{64}$$

These are Lamb's equations (28) and (29), p. 160, generalised.

Similarly his equation 30 generalised is

$$2T = \iiint D_1 S\rho\sigma_1\, \nabla_1 \sigma_1\, d\mathfrak{s}, \tag{65}$$

for

$$\iiint D_1 S\rho\sigma_1\, \nabla_1\sigma_1\, d\mathfrak{s} = \iint DS\rho\sigma\, d\Sigma\,\sigma - \iiint DS\,\zeta\sigma\zeta\sigma\, d\mathfrak{s},$$

by equation (9) § 6 above. But

$$\zeta\sigma\zeta = 2\zeta S\sigma\zeta - \zeta^2\sigma = \sigma,$$

and the surface integral vanishes as before.

Viscosity.

83. To consider viscosity the assumption is made that the shearing stress which causes it is $\mu \times$ the rate of shear of the moving fluid. μ is assumed independent of the velocity and experiment seems to shew that it is also independent of the density. This last we assume though the variation with density can be easily treated.

Consider a general strain $\chi\omega$. Since

$$mS\lambda\mu\nu = S\chi\lambda\chi\mu\chi\nu = S\lambda\chi' V\chi\mu\chi\nu^*,$$
$$V\chi\mu\chi\nu = m\chi'^{-1}V\mu\nu.$$

Putting μ and ν for any two vectors perpendicular to ω we see that the normal $U\omega$ to any interface becomes $U\chi'^{-1}\omega$ by the strain. Hence the interface ω experiences a shear (strain) which equals the resolved part of $\chi U\omega$ perpendicular to the vector $\chi'^{-1}\omega$ equals resolved part of $(\chi - \chi'^{-1})U\omega$ perpendicular to $\chi'^{-1}\omega$. Now when χ is the strain function due to a small displacement $\sigma\, dt$, by § 12 above

$$\chi\omega = \omega - S\omega\nabla \,.\, \sigma\, dt, \tag{66}$$

whence

$$\chi'^{-1}\omega = \omega + \nabla_1 S\omega\sigma_1\, dt, \tag{67}$$

and we see that the shear is the resolved part of

$$-(SU\omega\nabla \,.\, \sigma + \nabla_1 SU\omega\sigma_1)\, dt$$

perpendicular to ω, i.e. parallel to the interface ω. From this we see by our assumption concerning viscosity that any element of the fluid is subject to a stress ϕ given by

$$\phi\omega = -R\omega - \mu(S\omega\nabla \,.\, \sigma + \nabla_1 S\omega\sigma_1)$$

R being a scalar. Now we define the pressure p by putting

$$3p = S\zeta\phi\zeta = 3R + 2\mu S\,\nabla\sigma$$
$$\therefore \quad \phi\omega = -p\omega + \tfrac{2}{3}\mu\omega S\,\nabla\sigma - \mu(S\omega\nabla \,.\, \sigma + \nabla_1 S\omega\sigma_1). \tag{68}$$

The equation of motion is

$$D\dot{\sigma} = D\mathbf{F} + \phi\Delta,$$

or

$$D(\partial\sigma/\partial t - S\sigma\nabla \,.\, \sigma) = D\dot{\sigma} = D\mathbf{F} - \nabla p - \mu\nabla S\,\nabla\sigma/3 - \mu\nabla^2\sigma. \tag{69}$$

If μ be not as stated independent of D this equation must be made to contain certain space fluxes of μ which are quite easy to insert.

*Kelland and Tait's *Quaternions*, chap. x. equation (n). We have already used a particular case of this in § 15, above.

84. In § 179 of Lamb's treatise he considers the dissipation function due to the viscosity of a fluid in a way that seems to me misleading if not wrong. It appears as if he should add to the first expression of that section

$$u\,dp_{xx}/dx + v\,dp_{xy}/dx + w\,dp_{xz}/dx.$$

He refers to Stokes, *Camb. Trans.* vol. IX. p. 58. Let us give the quaternion treatment of Stokes's method.

If T is the kinetic energy of any portion of the fluid

$$2T = -\iiint D\sigma^2\,d\mathfrak{s}.$$

Thus

$$\because \quad d(D\,d\mathfrak{s})/dt = 0,$$

$$\dot{T} = -\iiint DS\,\sigma\dot{\sigma}\,d\mathfrak{s},$$

but

$$D\dot{\sigma} = D\mathbf{F} + \phi_1\nabla_1,$$

so that

$$\dot{T} = -\iiint DS\,\sigma\mathbf{F}\,d\mathfrak{s} - \iiint S\,\sigma\phi_1\nabla_1\,d\mathfrak{s}$$

or

$$\dot{T} = -\iiint DS\,\sigma\mathbf{F}\,d\mathfrak{s} - \iint S\,\sigma\phi\,d\Sigma + \iiint S\,\sigma_1\phi\nabla_1\,d\mathfrak{s}. \tag{70}$$

If now ϕ be expressed in terms of p and μ we have $\dot{T}$ depending on $\mathbf{F}$, p and μ. The part of $\dot{T}$ depending on $\mathbf{F}$ and p represents energy stored up as potential energy of position and potential energy of strain respectively, but the part depending on μ represents a loss of energy to the system we are considering the energy being converted into heat.

Thus putting $\phi = p + \varpi$ we have by equation (68)

$$\left.\begin{aligned}\varpi\omega &= \tfrac{2}{3}\mu\omega S\,\nabla\sigma - \mu(S\,\omega\nabla\,.\,\sigma + \nabla_1 S\,\omega\sigma_1)\\ &= \tfrac{2}{3}\mu\omega S\,\zeta\psi\zeta + 2\mu\psi\omega,\end{aligned}\right\} \tag{71}$$

where ψ is given by equation (75) below.

Thus we see that the rate of loss of energy is

$$\iint S\,\sigma\varpi\,d\Sigma - \iiint S\,\sigma_1\varpi\nabla_1\,d\mathfrak{s}. \tag{72}$$

The surface integral is the work done by viscosity against the moving fluid at the boundary and the volume integral is considered due to the work done against the straining of the fluid. Thus we put

$$F = -S\,\sigma_1\varpi\nabla_1, \tag{73}$$

and call F the "dissipation function."

By equation (6) § 3 above

$$F = -S\psi\zeta\varpi\zeta, \tag{74}$$

where ψ is the rate of pure strain of the fluid, i.e.

$$2\psi\omega = -S\omega\nabla \,.\, \sigma - \nabla_1 S\omega\sigma_1. \tag{75}$$

Again by equation (18) § 18 above because F is quadratic in ψ

$$F = -S\psi\zeta_\psi \daleth F\zeta/2, \tag{76}$$

so that from equation (74) we have, by § 4 above,

$$\varpi = {}_\psi\daleth F/2. \tag{77}$$

Substituting for ϖ from equation (71) in equation (74)

$$F = -\tfrac{2}{3}\mu(S\zeta\psi\zeta)^2 - 2\mu\psi\zeta\psi\zeta, \tag{78}$$

which gives F in terms of ψ.

SECTION VI.

THE VORTEX-ATOM THEORY.

85. If Quaternions can give valuable hints or indicate a promising method of dealing with the highly interesting mathematical theory of Vortex-Atoms, I think this alone ought to be sufficient defence of its claims to be within the range of practical methods of investigation.

In what follows I think I may be said to have indicated a hopeful path to follow in order to test to some extent the soundness of this theory.

Statement of Sir Wm. Thomson's and Prof. Hicks's theories.

86. Sir Wm. Thomson's theory is so well-known that it is not necessary to state it in detail. Matter is some differentiation of space which can vary its position carrying with it so to speak certain phenomena, some of which admit of definite quantitative measurement. Perhaps the most important of these phenomena is what is called mass. Now, says in effect Sir Wm. Thomson, if we suppose all space filled with an incompressible perfect fluid the vortices in it are just such differentiations of space. They also carry about with them definite quantitative phenomena. Can we prove that these hypothetical vortices would act upon one another as do the atoms of matter, the laws of whose action are contained in the various Sciences, e.g. Physics, Chemistry and Physiology? The problem in its first stages at any rate is a mathematical one, but during the many years it has been before the mathematical world very little progress has been made with it.

Hicks's extension of this theory is perhaps not so well known, but it seems to me quite as interesting and more likely to tally with the known phenomena of matter. He enunciates his theory in the *Proceedings of the Cambridge Philosophical Society*, vol. III. p. 276. It differs from Thomson's simply in assuming that the fluid does not quite fill space—that there are in it bubbles*. These bubbles will

*By "bubbles" of course I do not mean spaces occupied by another kind of fluid of smaller density but actual vacua. Thus a bubble may start into existence where none previously existed, or again a bubble may completely disappear.

find their way to where the pressure is least, i.e. speaking generally to the centre of some at least of the vortices. Thus we have another source of differentiation of space and general considerations seem to point to *these* differentiations being the true atoms, though of course "atom" is no longer a descriptive term.

General considerations concerning these theories.

87. In Maxwell's article *Atom* in the *Encyc. Brit.* p. 45, he says—"One of the first if not the very first desideratum in a complete theory of matter is to explain first mass and second gravitation...... In Thomson's theory the mass of bodies requires explanation. We have to explain the inertia of what is only a mode of motion and inertia is a property of matter, not of modes of motion. It is true that a vortex ring at any given instant has a definite momentum and a definite energy, but to shew that bodies built up of vortex rings would have such momentum and energy as we know them to have is in the present state of the theory a very difficult task.

"It may seem hard to say of an infant theory that it is bound to explain gravitation."

Now as Hicks tells us what induced him to give his theory was the promise it gave of explaining gravitation. But I believe nowhere does he point out the still more important result that probably on his theory we can explain inertia.

The statement of the principal property of inertia put scientifically is that the motion of the centre of gravity of any two bodies approximates more and more nearly to uniform velocity in a straight line, the more nearly they are isolated from external influence. But this property is probably true of Hicks's bubbles. The centre of gravity of any portion of the fluid containing certain bubbles (1), (2) ... (n) will if approximately isolated from all the rest of the fluid move approximately in a straight line, but this amounts to saying that the centre of volume of the n bubbles will also move uniformly in a straight line *if* the centre of the whole volume (bubbles *and* fluid) considered, move similarly. These conditions are not evidently true, but I think they are probably so when we consider groups of large numbers of bubbles.

Whether this theory explains gravitation is one of the principal questions to be considered in the following first trial.

Description of the method here adopted.

88. Finding it practically impossible to consider the real problem of a number of bubbles in the liquid, I consider the fluid continuous and not incompressible. Let us assume that

$$D = \tanh(p/c), \tag{1}$$

where c is some very small constant which in the limit = 0. For ordinary values of p, D very nearly = 1. It is only when p becomes comparable with c that D varies. When $p = 0$, $D = 0$. Also

$$P = \int(dp/D) = c \log \sinh(p/c). \tag{2}$$

For ordinary values of p then, $P = p$ but when p becomes comparable with c, P varies and when $p = 0$, $P = -\infty$. If we assume then for the greater part of the fluid that p is large compared with the small quantity c we see that the fluid will have almost exactly the same properties as a liquid containing bubbles. We may now apply the equations of motion of § 64, § 65.

Now we know the velocity at any point of an infinite fluid in terms of the spins and convergences at all points. From this we may deduce the acceleration in terms of the spins, convergences and time-fluxes. It so happens that by the equations in § 73 we can get rid of the time-fluxes of the spins. This greatly simplifies the further discussion of the problem. The equation that we thus deduce forms the starting point of our investigation, for the phenomena of gravitation, electro-magnetism, stress, &c., are exhibited and measured by the *acceleration* in bodies due to their relative positions.

The equation we obtain at once gives a generalisation of the integral, equation (12) § 65 above.

At the end of this section I give an equation which is rather complicated but which promises to enable us to deal more rigorously with our problem than I profess to have done below.

We proceed to the investigation just indicated.

Acceleration in terms of the convergences, their time-fluxes, and the spins.

89. Let us put for the convergence and twice the spin m and τ respectively, so that

$$S\nabla\sigma = m, \tag{3}$$

$$V\nabla\sigma = \tau. \tag{4}$$

We have seen that in equation (50) § 79 we may neglect the surface integral.

Thus

$$4\pi\sigma = \nabla \iiint u(m+\tau)\, d\mathfrak{s}, \qquad (5)$$

$$\therefore \qquad 4\pi\partial\sigma/\partial t = \nabla \iiint u\partial(m+\tau)/\partial t \,.\, d\mathfrak{s}.$$

By § 73

$$\dot{\tau} = \tau S\nabla\sigma - S\tau\nabla \,.\, \sigma,$$

but by equation (1) § 62

$$\partial\tau/\partial t = \dot{\tau} + S\sigma\nabla \,.\, \tau,$$

$$\therefore \qquad \partial\tau/\partial t = S\sigma\nabla \,.\, \tau + \tau S\nabla\sigma - S\tau\nabla \,.\, \sigma,$$

or because $S\nabla\tau = 0$

$$\partial\tau/\partial t = \tau S\Delta\sigma - \sigma S\Delta\tau = V\nabla V\sigma\tau \qquad (6)$$

$$\therefore \qquad 4\pi\partial\sigma/\partial t = \nabla \iiint u(V\nabla V\sigma\tau + \partial m/\partial t)\, d\mathfrak{s}. \qquad (7)$$

This equation is not in a convenient form for our purpose, for τ and m may be and most probably will be discontinuous, so that it is advisable to get rid of their derivatives. Moreover it is advisable to allow the time-flux of m to appear only under the form $d(m\, d\mathfrak{s})/dt$ because in case of pulsations the time integral of this expression will be zero, whereas we can predicate no such thing of $(\partial m/\partial t)\, d\mathfrak{s}$, or indeed of $\dot{m}\, d\mathfrak{s}$. Let us first then consider the first term in equation (7), viz. $\nabla \iiint uV\nabla V\sigma\tau\, d\mathfrak{s}$. Using equation (9) § 6, and neglecting the surface integral at infinity, this becomes

$$\begin{aligned} &-\nabla \iiint V\nabla u V\sigma\tau\, d\mathfrak{s} = \nabla V\nabla \iiint uV\sigma\tau\, d\mathfrak{s} \qquad [\S\, 9] \\ &= \nabla^2 \iiint uV\sigma\tau\, d\mathfrak{s} - \nabla S\nabla \iiint uV\sigma\tau\, d\mathfrak{s} \\ &= 4\pi V\sigma\tau + \nabla \iiint S\sigma\tau\nabla u\, d\mathfrak{s}, \end{aligned}$$

by equation (19) § 10 above.

Remembering now (§ 65 above), that

$$\dot{\sigma} = \partial\sigma/\partial t - V\sigma\tau - \nabla(\sigma^2)/2,$$

we see by equation (7) that

$$4\pi\dot{\sigma} = -2\pi\nabla \,.\, \sigma^2 + \nabla \iiint S\sigma\tau\nabla u\, d\mathfrak{s} + \nabla \iiint u(\partial m/\partial t)\, d\mathfrak{s}, \qquad (8)$$

whence by equation (9) § 65 above

$$P + v - \sigma^2/2 + (4\pi)^{-1} \iiint (S\sigma\tau\nabla u + u\,\partial m/\partial t)\, d\mathfrak{s} = H, \qquad (9)$$

where H is a function of the time only. This is a generalisation of equation (12) § 65, for assuming that $\tau = 0$ we know by § 79 that $\partial\phi/\partial t = (4\pi)^{-1} \iiint u(\partial m/\partial t)\, d\mathfrak{s}$.

90. This integral equation may be put into several different forms by means of equation (9) § 6 above. The form that is useful to us is the one in which instead of $\partial m/\partial t$ we have $d(m\,d\mathfrak{s})/dt$ as we have already seen. Now $d(d\mathfrak{s})/dt = -m\,d\mathfrak{s}$. Therefore

$$\begin{aligned} d(m\,d\mathfrak{s})/dt &= (\dot{m} - m^2)\,d\mathfrak{s} \\ &= (\partial m/\partial t - S\sigma\,\nabla m - m^2)\,d\mathfrak{s}. \end{aligned}$$

Substituting from this for $\partial m/\partial t$, and then transforming by equation (9) § 6 so as to get rid of the space variations of m involved in $S\sigma\,\nabla m$ we get

$$\begin{aligned} \iiint u\,(\partial m/\partial t)\,d\mathfrak{s} &= \iiint u\,d(m\,d\mathfrak{s})/dt + \iiint (um^2 - mS\sigma\,\nabla u - um^2)\,d\mathfrak{s} \\ &= \iiint u\,d(m\,d\mathfrak{s})/dt - \iiint mS\sigma\,\nabla u\,d\mathfrak{s}. \end{aligned}$$

Thus from equations (8) and (9)

$$4\pi\dot{\sigma} = -2\pi\nabla\,.\,\sigma^2 + \nabla\iiint S\,\nabla u(V\sigma\tau - m\sigma)\,d\mathfrak{s} + \nabla\iiint u\,d(m\,d\mathfrak{s})/dt, \qquad (10)$$

and

$$P + v - \sigma^2/2 + (4\pi)^{-1}\iiint\{d\mathfrak{s}\,S\,\nabla u(V\sigma\tau - m\sigma) + u\,d(m\,d\mathfrak{s})/dt\} = H. \qquad (11)$$

Sir Wm. Thomson's Theory.

91. We are now in a position to examine the two theories. We first take Thomson's, which is considerably the simpler, and which therefore serves as an introduction to the other. We have then $m = 0$. Thus equations (10) and (11) become

$$4\pi\dot{\sigma} = -2\pi\nabla\,.\,\sigma^2 + \nabla\iiint S\,\sigma\tau\,\nabla u\,d\mathfrak{s}, \qquad (12)$$

$$p/D - \sigma^2/2 + (4\pi)^{-1}\iiint S\,\sigma\tau\,\nabla u\,d\mathfrak{s} = H, \qquad (13)$$

for in the present theories we may put $v = 0$.

We shall consider the two terms on the right of equation (12) separately. The second term gives then an apparent force per unit mass due to a potential

$$-(4\pi)^{-1}\iiint S\,\sigma\tau\,\nabla u\,d\mathfrak{s}. \qquad (14)$$

Comparing this with equation (33) § 46 above we see that this potential is the same as that of a magnetic system given by

$$4\pi\mathbf{I} = V\sigma\tau. \qquad (15)$$

Now this magnetism is zero where τ is zero. In other words it is only present where the vortex-atoms are, and therefore it cannot be so distributed as to give an apparent force of gravitation, for taking the view of magnetic matter expressed in equation (34) § 46 we see that there must, in every complete vortex-atom, be a sum of magnetic matter exactly = 0. Nor again is it likely to explain the phenomena of permanent magnets, because, assuming that for any given small space including many vortex-atoms $\iiint V\sigma\tau\, d\mathfrak{s}$ is not zero, the apparent force produced will affect all other parts of space independently of whether this same integral for them is not or is zero. But to explain the phenomena of permanent magnets we must assume that the effect takes place only on portions of space where there is positive magnetic matter. This term then gives us no phenomena analogous to physical phenomena. As a matter of fact it probably has no visible effect on large groups of vortices, for there is no reason to suppose that the vector $V\sigma\tau$ is distributed otherwise than at random.

92. Let us now consider the other term in equation (12), and neglecting the term already considered, put

$$\dot{\sigma} = -\nabla \cdot \sigma^2/2. \tag{16}$$

The phenomena resulting from this are the same as would follow from a stress in a medium, the stress being an equal tension in all directions $= -\sigma^2/2$. Now comparing equation (62) § 58 with equation (4) § 89, we see that σ depends on $\tau/4\pi$ in exactly the same way as does **H** on **C**. The question then arises—is the stress we are now considering equivalent in its mechanical effects upon the vortex-atoms to the stress given by equation (68) § 60 above, which explains the mechanical effect of one current on another? We saw in § 60 that this stress is a tension $-\mathbf{H}^2/8\pi$ along the vector **H** and an equal pressure in all directions at right angles. The effects then would be the same only if σ be at right angles to the surface of our atom. But this is obviously not in general the case. From this analogy we can see however what approximately will happen to our atom. For instead of σ being at right angles to the surface it is in all probability very nearly tangential. Assuming that it is actually tangential we see that at the surface of the atom we have a tension exactly corresponding to the pressure which in the electric analogue will be exerted on this surface. In other words, the atoms will act on each other very approximately in what may be called a converse way to the small circuits in the electric analogue; i.e. where, in the electric analogue there is an attraction, in the hydrodynamic case there will be an apparent repulsion, and vice versâ.

Now each vortex-atom forms a small circuit and therefore acts in the converse way to a small magnet. In other words, each atom acts upon each other atom as if it were charged with *attracting* magnetic matter. Thus we see that if we could suppose certain extra atomic vortices to exist and to be disposed throughout space in what at present must be considered quite an artificial manner with reference to the *atomic* vortices we could rear up a fabric which would explain gravitation. This conception however is of very little use for our present purpose.

From these considerations I think we have every reason to believe that Thomson's theory in its native simplicity does not promise to lead us to the physical phenomena of matter. We pass on therefore to Hicks's.

Prof. Hicks's Theory.

93. Hicks in his theory, as I understand him, assumes that the bubbles always remain associated with the same particles of the fluid. This of course is probably not the case. By reason of the variation of the pressure with the time it is probable that evanescent bubbles start into existence and disappear at various parts of the fluid. This requires some few preliminary remarks.

The particles of the fluid with which bubbles are permanently (i.e. throughout the greater part of each small but not infinitely small interval of time) associated are those where the intensity of spin is greatest. If the intensity of spin is quite various at different points we shall thus have vortices where there is generally no bubble, extra material vortices in fact. We must suppose these distributed quite at random till the more exact mathematical treatment of our problem leads us to suppose otherwise. Now evanescent bubbles will occur rather in these vortices than in parts of the fluid where is no vortex at all (if we may suppose such parts to exist), and of course they will occur more readily in stronger than in weaker vortices. At the present stage of the theory then we may suppose evanescent bubbles to occur in all parts of the fluid. As a first approximation to the consideration of the effect of these bubbles we may assume a part m' of m to be continuously distributed through space. Putting

$$m = m' + M, \tag{17}$$

we must suppose M to be present only at the material bubbles where it is probably discontinuous, whereas m' is continuous throughout both the material vortices and the rest of the fluid.

Now when on account of variation of pressure m' and M are affected—is it probable that in the neighbourhood of a permanent bubble m' and M are of the same or opposite signs? To answer this question observe that what we call m' continuously distributed is really a series of discontinuous values of m scattered at random through space so that m' is probably very small. A decrement of pressure will cause an increment of evanescent bubbles, i.e. a decrement in m'. An increment of the permanent bubbles will also take place the magnitude of which by what has been said concerning m' will not be by any means accounted for by the decrement in m'. There will therefore also be a decrement in M. Similarly for an increment in the pressure. M and m' may therefore be assumed to be of the same sign.

After noticing that m' is continuous and therefore that there is no objection to introducing its space variations we are furnished with all the materials necessary for discussing our problem. The equation we shall use is (10) of § 90 above. We divide its discussion into two parts as follows.

Consideration of all the terms except $-\nabla\,.\,(\sigma^2)/2$.

94. The reasons that we have already seen in § 91 for neglecting $\nabla\iiint S\,\sigma\tau\,\nabla u\,d\mathfrak{s}$ still hold good so we put this aside. This is not the case with $-\nabla\iiint mS\,\sigma\,\nabla u\,d\mathfrak{s}$ but we can neglect $\nabla\iiint u\,d(m\,d\mathfrak{s})/dt$ for the average value of $d(m\,d\mathfrak{s})/dt$ for any particle is zero. The only term to consider then is $-\nabla\iiint mS\,\sigma\,\nabla u\,d\mathfrak{s}$. Putting as in equation (17) $m = m' + M$ we have

$$-\nabla\iiint MS\,\sigma\,\nabla u\,d\mathfrak{s} - \nabla\iiint m'S\,\sigma\,\nabla u\,d\mathfrak{s}.$$

The first term of this can be neglected for the same reasons as for neglecting that containing $V\sigma\tau$. Applying equation (9) § 6 to the last term and neglecting the surface integral as usual we get

$$\nabla\iiint uS\,\nabla(m'\sigma)\,d\mathfrak{s} = \nabla\iiint u\{m'(m' + M) + S\,\sigma\,\nabla m'\}\,d\mathfrak{s}. \qquad (18)$$

The last term can probably be neglected though we cannot give such good reasons as for the other terms we have neglected. At any rate in places not near permanent bubbles $S\sigma\nabla m'$ is as likely to be positive as negative and vice versâ, so that such portions of space will on the whole produce no effect on the permanent bubbles. If $S\,\sigma\,\nabla m'$ contributes anything for parts of space in the neighbourhood of permanent bubbles we must be content at present with the assumption either that the contribution is in general positive or that if it be negative it is not sufficient to cancel the effect of the positive term $m'M$. Remembering that m' is small compared with M we are left with the positive term $m'M$. This as can be easily seen from the form of equation (18) leads to an apparent law of gravitation for our permanent bubbles.

95. The gravitational mass which we must on this supposition assign to each permanent bubble varies as the average value of $m'M$ for that bubble*. Now we saw in § 87 that the probable measure of mass of a permanent bubble was proportional to its average size. Do these two results agree? I cannot say, but even if they do not these considerations would still explain the motions of the solar system, but if the sun and Jupiter (say) were to collide their subsequent motion would not be that due to the collision of two bodies the ratio of whose masses is that which is accepted as the ratio of the sun's and Jupiter's. As a matter of fact however I should imagine that the average value of $m'M$ for a permanent bubble is proportional to its average volume and this simply as a consequence of the reasoning in § 87 above.

A conclusion at any rate to be drawn from the above is that there is a *presumption* in favour of Hicks's theory explaining gravitation.

Consideration of the term $-\nabla \cdot (\sigma^2)/2$.

96. In considering this term we adopt the method of § 92 and consider an electric analogue. The analogue is an electro-magnetic field for which in the notation of § 46 to § 60 above at every point,

$$\mathbf{H} = \sigma. \tag{20}$$

For this field we have at once

$$4\pi\mathbf{C} = V\nabla\mathbf{H} = \tau. \tag{21}$$

The distribution of the magnetism in the field is somewhat arbitrary, but in the notation of equations (61) and (62) § 82 it will be found that everything is satisfied by putting

$$4\pi\mathbf{I} = -\nabla S q. \tag{22}$$

This gives as it should $S\nabla(\mathbf{H} + 4\pi\mathbf{I}) = 0$, which is in fact the only equation it is necessary to satisfy. We have further

$$\mathbf{B} = \mathbf{H} + 4\pi\mathbf{I} = \sigma - \nabla S q = \nabla V q,$$

so that

$$\mathbf{A} = Vq. \tag{23}$$

*There is one important difference to be noticed between this and Hicks's explanation of gravitation. His depends on the *synchronous* pulsations of distant vortices. I do not wish to imply that I do not believe in the existence of such synchronous pulsations, but by the above we see that gravitation can probably be explained independently of them. As a matter of fact such synchronous pulsations probably actually occur on account of the variation of H with the time.

Thus all the important vectors in the analogue are determined. It remains to compare the mechanical effects of the analogue with the term $-\nabla \, . \, \sigma^2/2$.

I, it must be observed, is not confined to the bubbles, but is distributed throughout space.

97. If we now assume that bubbles have not always existed in the positions which we call permanent, there cannot at the surface of the bubbles be any circulation round them. This makes the velocity at the surface almost normal to it, so that the stress given in equation (66) § 60 reduces to a tension $-S\,\mathbf{H}(2\mathbf{B}-\mathbf{H})/8\pi$ over the surface, i.e. we have a pressure

$$-\sigma^2/8\pi + S\sigma\,\nabla Vq/4\pi.$$

Now on account of the absence of circulation ∇Vq is very small and may therefore be neglected. Thus we get a *pressure* $-\sigma^2/8\pi$, and are thus led once more to a "converse" of the analogue. This at once* leads to another reason for the law of gravitation if the pulsations are synchronous. This we have already seen to be probable.

The present consideration of the subject is merely to point to a *method* of investigating the theory of vortex-atoms. I therefore leave the subject here, not attempting to force the phenomena we have been considering to tally with the known phenomena of electricity and magnetism. Nevertheless I may say that the prospect of discussing these things by means of the present subject can scarcely be considered as distant after what has gone before.

To sum up, this first application of the method leads to a presumption in favour of Hicks's theory leading to an explanation of both the important properties of matter—inertia and the law of gravitation—and there is also reason from it to hope that the phenomena of electro-magnetism are not unlikely to receive an explanation. Thomson's theory on the other hand would seem to fail in the first two at any rate of those endeavours.

98. We close the essay with the fulfilment of the promise made towards the end of § 88. In that section it will be remembered we considered a hypothetical fluid for which $D = \tanh(p/c)$, and made this do duty for a liquid containing bubbles. Strictly speaking our liquid is *bounded* at the bubbles and therefore as a bounded liquid should it be treated. For such a liquid we require an equation

*[Note added, 1892. Because the density of attracting magnetic matter of the analogue = $-S\,\nabla\mathbf{H}/4\pi = -m/4\pi$.]

corresponding to equation (10) § 90, and if possible equation (11) also. This last I have been unable to obtain, and I am not sure that to solve the problem explicitly is possible.

Our problem only deals with an incompressible fluid, but as the removal of this restriction does not greatly complicate the work we will consider the general case of a bounded compressible fluid. We have

$$\begin{aligned} 4\pi\sigma &= \nabla^2 \iiint u\sigma\, d\mathfrak{s} = -\nabla \iiint \nabla u\sigma\, d\mathfrak{s} \\ &= \nabla \iiint u(\tau + m)\, d\mathfrak{s} - \nabla \iint u\, d\Sigma\, \sigma, \end{aligned}$$

whence

$$4\pi\partial\sigma/\partial t = \nabla \iiint d\{u(\tau + m)\, d\mathfrak{s}\}/dt - \nabla \iint d(u\, d\Sigma\, \sigma)/dt.$$

The justification of using $\partial/\partial t$ on the left and d/dt under the integral sign will appear if the increment $(\partial\sigma/\partial t)\, dt$ in σ at a given point in the time dt be considered. It will be observed that the meaning here to be attached to $\dot{u}$ will be $-S\sigma\,\nabla u$, as in the differentiation $\partial/\partial t$ with regard to the time the origin of u is assumed to be fixed.

I shall now merely indicate the method of procedure. By the method exhibited in § 89 we can prove that

$$d(u\tau\, d\mathfrak{s})/dt = -V\nabla u\, V\sigma\tau\, d\mathfrak{s} - u\sigma S\,\tau\Delta\, d\mathfrak{s}$$

and that

$$d(um\, d\mathfrak{s})/dt = -mS\,\sigma\,\nabla u\, d\mathfrak{s} + u\, d(m\, d\mathfrak{s})/dt.$$

From this we can deduce that

$$\begin{aligned} 4\pi\,\partial\sigma/\partial t = {} & \nabla V\,\nabla \iiint uV\sigma\tau\, d\mathfrak{s} + \nabla \iiint \{-mS\,\sigma\,\nabla u\, d\mathfrak{s} + u\, d(m\, d\mathfrak{s})/dt\} \\ & - \nabla \iint u\sigma S\,\tau\, d\Sigma - \nabla \iint u\, d(d\Sigma\,\sigma)/dt + \nabla \iint d\Sigma\,\sigma S\,\sigma\,\nabla u, \end{aligned}$$

and from this again we get

$$\dot{\sigma} = \nabla w + \nabla w' + \nabla v, \tag{24}$$

where w and w' are scalars and v a vector given by

$$4\pi w = -2\pi\sigma^2 + \iiint \{d\mathfrak{s}\, S\,\nabla u(V\sigma\tau - m\sigma) + u\, d(m\, d\mathfrak{s})/dt\}, \tag{25}$$

so that w is in fact the $H - v - P$ of equation (11) § 90

$$4\pi w' = \iint (S\, d\Sigma\, \sigma S \sigma \nabla u - u\, dS\, d\Sigma\, \sigma/dt), \tag{26}$$
$$4\pi v = \iint (-u\sigma S \tau\, d\Sigma + V\, d\Sigma\, \sigma S \sigma \nabla u - u\, dV\, d\Sigma\, \sigma/dt). \tag{27}$$

This last equation may be put in what for our purposes is the more convenient form

$$4\pi v = -\iint \{u\sigma S \tau\, d\Sigma + d(uV\, d\Sigma\, \sigma)/dt\}. \tag{28}$$

Again it may be put in a form free from d/dt; for $V\, d\Sigma\, \dot{\sigma} = 0$ because the surface is a free surface and $d(d\Sigma)/dt = \nabla_1 S\, d\Sigma\, \sigma_1$*, as can easily be proved by considerations similar to those in § 83. Thus

$$4\pi v = \iint (-u\sigma S \tau\, d\Sigma + V\, d\Sigma\, \sigma S \sigma \nabla u - uV\nabla_1 \sigma S\, d\Sigma\, \sigma_1). \tag{29}$$

In the problem we have to discuss $m = 0$, so that w gives only terms which we discussed in considering Sir Wm. Thomson's theory. Observing that if in w we change $m\, d\mathfrak{s}$ into $-S\, d\Sigma\, \sigma$ we get for the part of w containing m, w'; we see that w' only gives terms that we have virtually discussed under Hicks's theory. We have however entirely neglected v. Are we justified in this? In the first place we have seen that if the bubbles have not always been associated with those parts of the fluid with which they now are there is round every bubble absolutely no circulation. This shows that for any one bubble $\iint V\, d\Sigma\, \sigma = 0$†, and therefore we are justified in neglecting the last term in equation (28). We are probably also

*[Note added, 1892. This should be $d(d\Sigma)/dt = -m\, d\Sigma + \nabla_1 S\, d\Sigma\, \sigma_1$, and therefore equation (29) should be

$$4\pi v = \iint \{V\, d\Sigma\, \sigma S \sigma \nabla u - u(\sigma S \tau\, d\Sigma - mV\, d\Sigma\, \sigma + V\nabla_1 \sigma S\, d\Sigma \sigma_1)\}.$$

This does not affect our present problem because $m = 0$ in our case.]

†One way out of many of proving this is as follows. Divide the bubble up into a number of infinitely near sections by planes perpendicular to the unit vector α. For any one section $\int S\, d\rho\, \sigma = 0$. Consider $d\Sigma$ to be the element of the surface cut off by the following four planes, (1) the plane of section considered, (2) the consecutive plane of section, (3) the two planes perpendicular to $d\rho$, and through the extremities of the element $d\rho$. Thus if x be the distance between the two sections, $d\rho = x^{-1} V\alpha\, d\Sigma$, whence

$$\iint S\, \alpha\, d\Sigma\, \sigma = 0,$$

for the surface of the bubble between the two sections. But adding, we may suppose this integral to extend over the whole bubble. Thus $S\alpha \iint V\, d\Sigma\, \sigma = 0$ for the whole bubble; therefore α being a quite arbitrary unit vector we have for the whole bubble $\iint V\, d\Sigma\, \sigma = 0$.

justified in neglecting the first term, for probably τ is very nearly tangential to the surface and therefore $S\tau\, d\Sigma = 0$.

[Note added, 1892. The whole of this last section is in rather a nebulous stage, and since writing it I have not had sufficient leisure to return to the matter. I hesitated whether to include it in the present issue. But since, notwithstanding the absence of any reliable results, it serves very well to illustrate how investigations are conducted by Quaternions, I have thought it worth publishing.

Should anybody feel inclined to attempt to apply the method or an analogous one it is well to note that in the *Phil. Mag.* June, 1892, p. 490, I have given the more general result sought in this last section. As the result is not there proved I give one proof which seems instructive. It exhibits the great variety of suitable quaternion methods of dealing with physical questions. It furnishes incidentally a fourth quaternion proof of the properties of vortices. It also illustrates how special quaternion methods developed for use in one branch of Physics at once prove themselves useful in other branches.

Adopting the notation and terminology of *Phil. Trans.* 1892, p. 686, §§ 5–7, let σ and τ be taken as an intensity and flux respectively, τ still being $= V\nabla\sigma$ and therefore $\tau' = V\nabla'\sigma'$. Thus σ' and τ' are the actual velocity and double spin respectively and the equation of motion is

$$\dot{\sigma}' = -\nabla'(v + P).$$

Putting $\sigma' = \chi'^{-1}\sigma$ and operating on the equation by χ'

$$\dot{\sigma} + \chi' d\left(\chi'^{-1}\right)/dt \,.\, \sigma = -\nabla(v + P).$$

Now

$$\chi' d(\chi'^{-1})/dt \,.\, \sigma = -\dot{\chi}'\chi'^{-1}\sigma = -\dot{\chi}'\sigma' = \nabla_1 S\sigma_1'\sigma' = \nabla \,.\, \sigma'^2/2.$$

$$\therefore \qquad \dot{\sigma} + \nabla \,.\, \sigma'^2/2 = -\nabla(v + P), \qquad (A)$$

which can easily be deduced from or utilised to prove Lord Kelvin's theorem concerning "flow," equation (23) § 72 above.

As d/dt is commutative with ∇, $\dot{\tau} = 0$ or τ is an absolute constant for each element of matter. This being interpreted at once gives the well-known properties of vortices in their usual form.

If in the equation $4\pi(v + P) = S \,.\, \nabla^2 \iiint u(v + P)\, d\varsigma$ we carry one ∇ across the integral sign, get rid of its differentiations which affect u by equation (9) § 6 above, and then do the same with the other ∇ we get

$$4\pi(v + P) = \iint \{(v + P)S\, d\Sigma\, \nabla u - uS\, d\Sigma\, \nabla(v + P)\} + \iiint u\nabla^2(v + P)\, d\varsigma.$$

At surfaces of discontinuity in σ, v and P will both be continuous, so that instead of $\iint (v+P)S\,d\Sigma\,\nabla u$ we may write $\iint_b (v+P)S\,d\Sigma\,\nabla u$. In the last equation substitute throughout for $\nabla(v+P)$ from equation (A). Thus

$$\begin{aligned} 4\pi(v+P) &= \iint_b (v+P)S\,d\Sigma\,\nabla u + \iint uS\,d\Sigma\,\dot\sigma - \iiint uS\,\nabla\dot\sigma\,d\mathfrak{s} \\ &\qquad + \left(\iint uS\,d\Sigma\,\nabla\,.\,\sigma'^2/2 - \iiint u\nabla^2\sigma'^2\,d\mathfrak{s}/2\right) \\ &= \iint_b (v+P)S\,d\Sigma\,\nabla u + \iint uS\,d\Sigma\,\dot\sigma - \iiint uS\,\nabla\dot\sigma\,d\mathfrak{s} + \iiint S\,\nabla u\,\nabla\,.\,(\sigma'^2)\,d\mathfrak{s}/2, \end{aligned}$$

which is equation (36) of the *Phil. Mag.* paper.

If the standard position and present position of matter coincide it is quite easy to prove that

$$\begin{aligned} dS\,d\Sigma'\,\sigma'/dt &= S\,d\Sigma(\dot\sigma + V\nabla_1\sigma\sigma_1) \\ d(S\,\nabla'\sigma'\,d\mathfrak{s}')/dt &= S\nabla(\dot\sigma + V\nabla_1\sigma\sigma_1)\,d\mathfrak{s}. \end{aligned}$$

Substituting for $S\,d\Sigma\,\dot\sigma$, $S\,\nabla\dot\sigma$ from these in the last equation we get equation (32) of the *Phil. Mag.* paper; but this equation can also be proved directly. It should be noticed that equations (34) and (35) of that paper have been wrongly written down from equation (32). In each read $+\nabla\,.\,(\sigma^2/2)$ for $-\nabla\,.\,(\sigma^2/2)$.]